Oculomotor Systems and Perception

Understanding visual experience has long challenged the best of human minds, from the Ancient Greeks' interest in optics to the study of visual perception by contemporary psychologists and neuroscientists. Today's scientific study of perception seeks to understand the nature of our experience in terms of the underlying mechanisms by which it occurs.

This text is the first to emphasize the role of oculomotor systems in perception. Oculomotor systems that regulate eye movements play an important role in accounting for certain qualities of visual experience. These systems are implicated in a wide array of perceptual topics, from apparent size, depth, and distance to apparent slant and vertical orientation. The text begins with a brief introduction to the basic characteristics of such oculomotor systems as those controlling vergence, pursuit, the vestibulo-ocular response, and saccadic eye movements. Also introduced are fundamental concepts in physiological optics. Next explored are mechanisms of perception, with a particular focus on eye movements, and the remarkably diverse implications of oculomotor research, which extend to motion sickness and life in space orbit. Insights into dysfunctional vision are offered as well.

Oculomotor Systems and Perception complements standard texts on visual perception, yet may be read independently by those with a modest background in vision science.

Sheldon M. Ebenholtz is Distinguished Professor Emeritus at the State University of New York College of Optometry and is an adjunct professor in the Department of Psychology at the University of Arizona. email: jme-sme@msn.com

Oculomotor Systems and Perception

Sheldon M. Ebenholtz

State University of New York, College of Optometry
University of Arizona, Department of Psychology

Sheldon M. Ebenholtz

State University of New York, College of Optometry
University of Arizona, Department of Psychology

CAMBRIDGE
UNIVERSITY PRESS

CAMBRIDGE UNIVERSITY PRESS
Cambridge, New York, Melbourne, Madrid, Cape Town, Singapore, São Paulo

Cambridge University Press
The Edinburgh Building, Cambridge CB2 2RU, UK

Published in the United States of America by Cambridge University Press, New York

www.cambridge.org
Information on this title: www.cambridge.org/9780521804592

First published 2001
This digitally printed first paperback version 2005

A catalogue record for this publication is available from the British Library

Library of Congress Cataloguing in Publication data
Ebenholtz, Sheldon M. (Sheldon Marshall), 1932–
    Oculomotor systems and perception / Sheldon M. Ebenholtz.
        p.   cm.
    Includes bibliographical references and index.
    ISBN 0-521-80459-0
    1. Eye – Movements.   2. Visual perception.   I. Title.
QP477.5 .E246   2001
612.8′46 – dc21   2001025114

ISBN-13 978-0-521-80459-2 hardback
ISBN-10 0-521-80459-0 hardback

ISBN-13 978-0-521-00236-3 paperback
ISBN-10 0-521-00236-2 paperback

This work is dedicated to my dear wife, Jean, whose energy and enthusiasm never once veered toward entropy, and for whose support I shall always be grateful; and to my remarkable son, Keith, who taught me fathering and the law.

I think a teacher's role should be limited to clearly showing his pupil the goal that a science sets itself and to pointing out all possible means at his disposal for reaching it. But a teacher should then leave his pupil free to move about in his own way and, according to his own nature, to reach his goal, only coming to his aid if he sees that he is going astray. I believe, in a word, that the true scientific method confines the mind without suffocating it, leaves it as far as possible face to face with itself, and guides it, while respecting the creative originality and the spontaneity which are its most precious qualities.

—Claude Bernard (1865/1957)
An Introduction to the Study
of Experimental Medicine

Contents

Figure Captions

Table Headings

Foreword

This work is intended to complement a standard text on visual perception with knowledge of the essential features of oculomotor systems and their bearing on visual perception. These issues, however, have useful implications for students in a number of disciplines such as psychology, biomedical engineering, human factors, optometry, ophthalmology, and artificial intelligence. Accordingly, the work is structured to be complete in itself and may be read on its own. Students not trained in optics or the vision-health disciplines also may find useful the Appendix on common visual anomalies as well as Chapter 2 on central concepts of physiological optics.

I have tried, throughout, to be mindful of historical and philosophy of science issues so that the myriad facts associated with the intersection of perception and eye movement systems may derive extended meaning by being seen in an appropriately broad context. My hope, of course, is that in so doing, reading the text proves to be both a useful and an enjoyable intellectual exercise.

S. M. E.
Goodyear, AZ

Preface

The theme of this book developed gradually as my teaching of perception evolved in courses and seminars, first at Connecticut College, in New London, and later at that remarkable research institution, the University of Wisconsin, Madison. My belief in the importance of the search for underlying mechanism in the explication of psychological phenomena, however, began still earlier at the New School for Social Research with my teachers Irvin Rock, Hans Wallach, Solomon Asch, and Mary Henle, and my peer group, Bill Epstein, Lloyd Kaufman, Martin Lindauer, and Carl Zuckerman. It was a wonderfully stimulating zeitgeist where phenomenology met the firm constraints of empiricism and where the democratic ethos worked its way into the classrooms so that challenges to current ideology were expected and encouraged.

Hopefully this attitude was not lost on my past graduate students, many of whom, like Judith Callan, Paul Dubois, S. K. Fisher, Ken Paap, and Wayne Shebilske, contributed to the empirical corpus of evidence that has helped to solidify the view that oculomotor systems underlie various aspects of perception, and thereby provide a theoretical alternative or at least a necessary complement to cognitive and computational explanations. And to all of my former students, including Tim Babler, Terry Benzschawel, Karl Citek, Changmin Duan, Gerry Glaser, Gordon Redding, Martin Steinbach, Mike Streibel, and John Utrie, who have distinguished themselves in all manner of endeavor, I wish to express my genuine appreciation for their curiosity, and above all for their commitment to learn the ways of truth seeking in science.

To Karl Citek, S. K. Fisher, Gerry Glaser, Wayne Shebilske, and Bob Welch, I am especially grateful for their helpful comments on various portions of this work.

A note of appreciation also must be expressed in connection with our scientific agencies and foundations, NIH, NIMH, and NSF. I am grateful not just for having survived many of their peer reviews, but for the positive role these government agencies have played in stimulating the commerce of science by providing the palpable means by which ideas may enter the public domain initially as testable hypotheses and ultimately providing the facts of our discipline.

Special mention is reserved for my colleagues Mal Cohen, Dick Held, Bob Kennedy, Len Matin, and Bob Welch, who over the years have provided both collegiality and much insight. Always remarkably stimulating and productive were my several extended visits to Mal's lab at the NASA–Ames Research Center, and to Bob Kennedy's lab in Orlando and to his utterly unique Pensacola–Ariola conferences in Pensacola Beach. A debt of gratitude also is owed to my colleagues and friends Marcia Ozier and Bruce Earhard, whose nomination led to my appointment as a Killam Senior Fellow during 1971–1972 at Dalhousie University in Nova Scotia. The idea that a change in the oculomotor vergence resting level was the mechanism underlying so-called distance adaptation was germinated during that remarkable year.

I am also pleased to take the opportunity of this public forum to thank the many colleagues who, over the years, shared their enthusiasm for new contributions to knowledge at various conferences, and at meetings of the Psychonomic Society and the Association for Research in Vision and Ophthalmology. Many of them grew into an extended family that nurtured mutual respect and valued scientific accomplishment. Finally, I offer a special note of appreciation, for the many opportunities to share time with Herschel and Eileen Leibowitz, and for their enduring legacy of civility and creativity.

1

Introduction

The Context for Perception

The scientific study of perception is the study of the qualities of experience and the conditions under which they occur. Although Gestalt psychologist Kurt Koffka (1935) set the scientific goal of explaining why the world *looks* as it does, his treatment of many other dimensions of consciousness, ranging from sound localization to cognition, indicated that there is in principle no reason why perception scientists should not also study all forms of experience, including the aesthetic experiences as well as pain. In fact, the discipline of perception is as broad as the states and varieties of consciousness itself.

Nevertheless, there are several reasons why the scientific study of visual experience preceded, and also seems to predominate, the study of other modalities of consciousness. First is the rule of scholarly inertia. Vision and visual perception have been studied for more than 2500 years (Wade, 1998), as a result of the Ancient Greeks' interests in astronomy and optics and to the subsequent realization that the eye could be treated as an optical instrument (Boring, 1950). Therefore, it continues to be studied simply because it has proved itself to be a valid and significant body of knowledge. Of course, the Ancient Greeks did not discover science or the empirical approaches to knowledge that require nature to be not just observed and thought about but also carefully manipulated in controlled environments: "the nature of things betrays itself more by means of the operations of art than when at perfect liberty" (Bacon, 1620, p. 341). It also has been the case that along with the development of the scientific approaches to knowledge, from the fourteenth century on, has come the questioning of current dogma in an ever-widening

1

domain of scientific progress. Accordingly, science creates its own growth in what may be a never-ending spiral of questioning, hypothesis building, testing, and discovery.

Because perception cannot be divorced from consciousness, it is appropriate that some of the major philosophic and metascientific issues surrounding the study of perception be discussed, albeit briefly, in order to provide the reader with a broad perspective for the perceptual phenomena covered in later chapters. The view offered of the nature of experience, a type of emergent dualism (Searle, 1992; Chalmers, 1995; Scott, 1995), is meant to allow the reader to place perception in the context of the mind–brain problem and to view the latter in a contemporary philosophic context. In what follows, however, no attempt is made to develop a philosophic tract on epistemology or ontology. Rather, the views expressed simply represent a set of metascientific precepts or at best, heuristics about mind and matter that are consistent with contemporary physical and neuro-science. For example, there is now, at the beginning of the twenty-first century, overwhelming evidence that brain function and consciousness are intimately related. Therefore, this fact is recognized in the discussion of the relation between structure and function, and of conscious states as one of two aspects of reality. The pro-intuitive presumption that conscious states are *real* serves, of course, to sidestep the various philosophic positions that challenge the special nature of consciousness in an otherwise physical universe. However, I believe it is scientifically fruitless to deny the special status of conscious states for, among other implications, to do so would be to foreclose on framing what may be the most significant scientific question of all time; namely, What is the nature of the physical basis for consciousness? In this spirit, in the following sections a few additional issues are raised about the nature of perceptual experience in the hope that the metascience of one era will yield the scientific subject matter of another.

Elements of Consciousness

The Illusion of Publicity

Apart from a historical perspective, the study of visual attributes such as color, size, depth, movement, and so on may have received priority among scientists because of certain phenomenologic characteristics that give them the appearance of being objective and publicly observable, and therefore as being legitimate objects of serious scientific inquiry. Certainly, in comparison with pain qualities and other emotive and aesthetic experiences, visual qualities are perceived to exist at some distance from and independently of the observer. In fact, the externalization of experience itself may be considered a scientific problem (von Bekesy, 1967, pp. 220–228), such that once the conditions of externalization are understood the inner–outer locus of experience could, in principle, be manipulated at will. Considerable progress in this direction occurred in the case of sound localization, where tones may be made to appear either within the head or at some distance in surrounding space (Hartmann, 1999). However, pain, in contrast to visual and auditory experience, may be the paradigm case for a subjective experience because pains always reside on or within the observer and therefore are not experienced as publicly observable. However, because all experience has its seat in the brain, including the quality of being at a distance, none can be said to be more objective than any other. It seems clear why naive realism, the reification of subjective events, is no longer held as a viable doctrine, although the sense of object permanence and stability that the experience of externalization provides is quite real. Therefore, another reason must exist for the preference toward visual problems.

Stimulation and Common Sensory Response

Although it is true of all the qualities of experience that they are private, and therefore that one can only *have* the experience but not observe it, nevertheless it is much easier to communicate about visual sensations than about pain or aesthetics. There is one overriding reason for this to be so. In order to communicate about blueness or many other visual qualities, all one must do is present the stimulation known to produce the experience in question. Even the ancients could do this using pigments, well before any correct formulation of the nature of radiation was available. As long as our nervous systems are not too different (see discussion of structure and function the following), then we can reasonably be assured that exposure of our sensory end organs to common stimulation evokes in each of us a similar though equally private experience. Thus, the scientific requirement of control and reliability of visual experience can be achieved relatively easily by manipulating the sources of stimulation. It seems quite likely, therefore, that the relative ease of controlling visual stimulation has been the major factor that has lent itself to the rapid development of the visual sciences. Let us now, by contrast, consider developing a similar approach to pain.

Ethical issues aside, the control of pain stimulation is quite complicated: many stimuli are internal and are not easily accessible, there is no specialized sensory end organ or common site of application, there is no readily available lexicon to describe the results of stimulation, and it is difficult to quantify the applied stimulation. It is apparent that the difficulties with the development of pain research lie not in the private nature of the pain experienced but in the control and application of the relevant stimulating and inhibiting conditions (Hilgard, 1978; Watkins & Mayer, 1985).

The study of aesthetic experience, likewise, would benefit from appropriate analysis and control of stimulating conditions. Of course, cognition and memory are of critical importance in this domain because they contribute to the stimulating conditions in the brain circuitry involved in the aesthetic experience. It seems reasonable to conclude that conscious experience varies from person to person to the extent to which idiosyncratic cognitive factors contribute to the stimulus package.

Aspects of Reality

Direct and Indirect Aspects

It is perhaps worth noting that with the exception of recursive brain circuitry, the stimulating conditions referred to previously represent the physical properties of our universe that are capable of eliciting a sensory response. These are the collections of atoms, molecules, photons, and all the radiation of the electromagnetic spectrum to which our receptors are responsive and which underlie the objects and qualities of our perceptual environment. These energy distributions constitute the physical aspect of reality that because of the fluid nature of science that proceeds by ruling out competing hypotheses, is always to be regarded as provisional and indirect. We get to know this aspect of reality only in terms of the most contemporary physical science concepts that have survived the rigorous procedures of scientific test and confirmation. Because physical reality is always formulated through the filter of science, it necessarily must be understood as provisional and indirect.

All the various types of energy distributions applied to receptors at the proper energy levels become transduced into a single type of electrochemical energy, which, in turn, is transmitted over neurons to the brain in the form of a train of nerve impulses (Tamar, 1972). One way the brain responds to this stimulation is with the production of conscious experience, the variety of which depends both on the temporal pattern of nerve impulses as well as on the particular spatial locus of stimulation received within the cortex. Conscious states and the various qualities of experience by which they are expressed represent the second aspect of reality, one that arises directly out of brain stimulation rather than through the inferential structure of science. Accordingly, indirect realities represent the physical entities (i.e., energy distributions) that may be said to exist in nature independently of any nervous system. In contrast, the direct realities that arise out of brain stimulation exist solely by virtue of their dependence on brain function. The qualities of sensory experience and other states of consciousness, thus, have no existence apart from a properly structured and stimulated brain from which they emerge. Much like two variables that when applied jointly in an experiment may produce a result that is more than the sum of the effects of each applied separately, consciousness itself may be regarded as an emergent effect of a nonlinear interaction (Scott, 1995). Also worthy of consideration is Sperry's (1985) suggestion that "the events of inner experience, as emerging properties of brain processes, become themselves explanatory causal constructs in their own right"(p. 379). Such a view actually finds expression in theoretical claims about perceptions causing perceptions such as increased perceived distance causes increased perceived size and so on. See Chapter 6 and also Kaufman (1974, p. 23) for a brief discussion of this issue.

Virtual Reality and Nonveridical Experience

Not all conscious states are in correspondence with certain energy distributions that constitute the physical objects of our environment. Dreams, hallucinations, and experience resulting from direct electrical brain stimulation (Penfield & Perot, 1963) represent conscious states produced without the presence of the physical energy distributions that normally would provide brain stimulation via peripheral sensory channels. When perceptual phenomena no longer justify a claim about the existence of some physical entity, then our perceptual experience cannot be said to be veridical or "truthful." This state of affairs occurs regularly in the course of illusions, the most common of which are due to the reflection of light from mirrors and movie screens. To be sure, although visual experience entailed by viewing the light reflected from a mirror or screen would of course be real, any claim for the existence of a physical object at the place signaled by the light rays would be in error. In general, experience may be nonveridical whenever we use the senses as though they were sensors of one or another energy distribution. Thus virtual reality devices warrant their name because no claim for the presence of the physical entities portrayed would be justified, although it would be proper to claim the existence of radiation in the visible range of wavelengths. However, drug-induced visual hallucinations would not even justify that claim and therefore may be considered totally nonveridical.

Brain as Mind Machine

The Relation of Structure to Function

There is one truly central problem that confronts all who have ever pondered the nature of perception, and that concerns the relation of the qualities of experience to the cortical structures that support them. Although this is, of course, an ancient question, there is a peculiarly modern ethical side to the mind–body issue that derives from an analysis of the relation between structure and function.

The brain is a remarkably complicated structure of billions of interconnecting neurons, some portion of which is associated with consciousness. In fact, the brain may be regarded as a mind machine, a device whose functions include the production of conscious states. Viewed in this fashion, like any other functional mechanism, brains represent mechanisms whose structures are intimately related to their respective functions. This is not to say that the same result, consciousness, may not occur from the exercise of different brain-building principles and different mechanisms or structures. Rather, it is to assert that if one builds two devices identically, except for trivial differences, then when activated, the functions or output of those devices must also be identical. When applied to brains, this has been called the Principle of Organizational Invariance, meaning that "any two systems with the same fine-grained *functional organization* will have qualitatively identical experiences" (Chalmers, 1995, p. 214). Because functional organizations may in principle be duplicated in computer chip architecture and biological systems, the interesting implication may be drawn that, according to this view, "the qualitative nature of experience is not dependent on any particular physical embodiment" (G. Glaser, personal communication, August, 2000).

Just what there is about the architecture and dynamics of brain processes, such as perhaps highly correlated streams of neuronal activity (Tononi & Edelman, 1998), that supports consciousness and accounts for the differences in sensory modalities, such as vision and hearing, is of course presently unknown. Nevertheless, neuroscientists are moving rapidly to develop the capability of simulating extant neural networks. For example, certain laboratories have grown neural networks on transistors with which the neurons have subsequently interacted (Service, 1999). Given the intimate relationship between structure and function, the ethical problem of having inadvertently stimulated consciousness in a simulated brain must then also be seriously considered.

A Role for Motor Systems in the Study of Perception

In the present work, only a restricted range of conscious experience is treated. Namely, those phenomena traditionally studied under the category of space perception. This includes the qualities of motion, depth, size, distance, slant, and spatial location. In the quest for scientific explanation of these phenomena, it has been useful to specify their antecedent conditions by examining the nature of stimulation, either in the sensory end organs (e.g., the retina) or in the brain itself. For example, motion perception may be determined by a certain succession of images across the retina, as occurs when viewing movies, or by a truly moving image. In either case, typical explanation is based on analysis of the properties of the pattern of stimulation input to the nervous system and associated internal processing channels (Livingstone & Hubel, 1987). In essence, this approach seems to imply that an explanation of sensory experience requires the explication of information input to the sensory channels as well as analysis of the structure of the channels themselves. Although it seems to be an intuitively proper heuristic to conduct perception research in this fashion, the approach, although fruitful, is necessarily incomplete. One reason for this is simply that motor systems, especially oculomotor systems, contribute certain qualities of experience, and no amount of analysis of sensory channels or sensory information can uncover the role played by motor systems.

From a historical perspective, the impetus for the separate study of sensory and motor systems stems, in the modern period, from the work of Charles Bell (1811) and Francois Magendie (1822). The Bell–Magendie Law, embodying their experimental results, ascribes a sensory function to the posterior roots of the spinal cord and a motor function to the anterior roots. This clear separation of sensory and motor function currently is recognized as efference, or nervous outflow from the brain; and afference, or inflow from sense receptors. Such a dichotomous representation, however, could not survive for long for two reasons. First, by the mid-1800s, various detectors [e.g., muscle spindles and Pacinian corpuscles, which signal muscle stretch and changes in cutaneous and deep pressure, respectively (Tamar, 1972)] had been found in muscle tissue and along cutaneous nerves and internal organs (Boring, 1942). Actually, it would appear that purely motor systems without a sensory apparatus may not exist as such. Second was the well-known concept of "innervation feelings" (Scheerer, 1987), culminating in Helmholtz's (1910/1962, Vol. 3, p. 245) proposal that efference, or the effort of will, may be registered in consciousness so that even without sensory feedback from peripheral detectors, awareness of limb, or eye position, would be possible. Thus, even at the origins of modern physiology and psychology, motor systems were thought to be implicated in space perception, and, conversely, Bell (1826) noted the importance of the "nervous circle," or what we would now call "sensory feedback," for precision of movement.

Presently, although there is a growing interest in the interactions between sensory and motor systems and their impact on perception (e.g., Bouwhuis et al., 1987), nevertheless, the specific role of oculomotor systems in perception is not widely known. In contemporary times, the tendency to treat sensory but not motor systems as sources of spatial information can be found even in developments in the robot-design community, where, for example, sophisticated electronic circuitry and machine-vision algorithms are used to drive robot motor systems (Indiveri & Douglas, 2000). Such modern endeavors are strangely reminiscent of seventeenth- and eighteenth-century approaches that, after Descartes' earlier mechanistic physiology, modeled behavior as a reflexive response to sensory processing (Peters, 1965), but ignored both the qualitative and metric contributions of the motor system itself to visual experience.

Egocentrically Speaking

Perhaps the most significant illustration of how an oculomotor system may play a role in perception may be drawn from the experience of egocentric direction. Any visually represented object is perceived in a space around an observer at some particular location in relation to the self, or ego. This, of course, is a remarkably functional fact, for knowing the position of an object enables one to act with respect to that object by, for example, catching or grabbing it, moving toward or away from it, or scanning its features. However, what conditions enable the observer to experience egocentric orientation to begin with?

If, like the owl, eyeball mobility in the head was restricted to only about one degree of arc (Steinbach & Money, 1973), then when an image was formed on the fovea, the object would invariably be located in front of the observer's head. So, if the position of the head on the trunk were known, then hands and feet (or wings) could be directed to move accordingly. However, because our eyes are quite mobile in the head, a foveally represented image, or an image represented anywhere else on the retina, may have come from an object at any one of a large number of places in space. Under these circumstances, there would no longer be a one-to-one relationship between the position of an image on the retina and the position of the head in relation to the object. Because a foveal image could arise from the light of an object above, below, left, or right of the observer, all depending on eye position, one is compelled to conclude that eye-position information must be present to disambiguate what would otherwise be an extraordinarily ambiguous location-finding task. Thus, oculomotor systems may be said to contribute a sense of "thereness" to visually apprehended objects.

Single points seen in otherwise total darkness are readily localized (Matin, 1986). Therefore, the quality of experience represented by "there," or ego-centric spatial location, as opposed to "what" (Ingle, Schneider, Trevarthen & Held; 1967, Held; 1968, Leibowitz & Post, 1982), is not necessarily based on the extraction of information in the scene from the optic array, as might be said of the perception of boundaries or contours. Rather, like the experience of color, it simply is attributed to the object based on extraretinal eye-position information (Matin, 1986). Accordingly, the role of oculomotor systems in imputing certain qualities of experience to objects is discussed in Chapter 4 as an *attributional* approach to perception.

The principal message of this volume is that oculomotor systems play a significant role in accounting for certain qualities of visual experience. No attempt is made, however, to address the classic issue of how the various oculomotor systems come to signal egocentric location or other visual attributes, whether through learning and development (Held & Hein, 1963; Hein & Diamond, 1983) or via evolutionary mechanisms (Rose, 1999). Once the importance of oculomotor systems has been stipulated, the need to study eye movement systems is apparent, and these are covered in Chapter 3. Furthermore, because eyes are mechano-optic systems, the need to introduce basic concepts in physical and physiological optics also follows directly. These matters are addressed in Chapter 2, whereas the Appendix introduces some common clinical problems that occur when the physiologic systems break down. Chapters 4 and 5 represent the critical mass of this work, with the empirical substrate for the main thesis in Chapter 4 and, a discussion of selected theoretical issues is left for Chapter 5. A succinct summary and a set of major unresolved problems associated with the present approach is contained in Chapter 6.

2

Some Basic Concepts of Physiological Optics

Introduction

The highly distilled knowledge represented in this chapter took centuries for the early Greek and later Arabic scholars to develop. Enormous efforts were made especially during the middle ages to transmit this knowledge from the Ancients by providing translations from Greek into Arabic and Latin, and later into the other languages of Western Europe (Lindberg, 1978a).

The issues that had to be argued about, thought through, and developed for about two millennia were extraordinarily basic, such as the rectilinear propagation of light; the structure and position in the eye of the sentient surface leading to perception; the nature of refraction, especially within the eye; vision due to extramission of radiation from the eye vs. intromission of light from objects into the eye; that light emanates in all directions from each point on an object; and that the relationships between points on an object had to be maintained in the image within the eye. It was not until 1583 that Felix Platter, a medical peer of Johannes Kepler, correctly placed the visual sensory mechanism in the retina and not at the lens (Lindberg, 1976, 191–208). Armed with this insight, in 1604 Kepler provided the correct refractive path of light through the cornea, pupil, lens, and all refracting media; offered the correct theory of the inverted, reversed retinal image; and demonstrated how all light rays emanating from a point in the visual field are brought to a focus at one point on the retina (Lindberg, 1976, 1978b).

In the context of oculomotor systems, it is recognized that visual stimulation plays a large, but not unique, role in the control of eye movements. As a step toward understanding the role of vision in oculomotor control, it is necessary to consider some of the basic concepts of physiological optics. These are covered in six topic headings:

Images
Light vergence
Lenses
Why lenses and prisms change the path of light
Prism diopters and ocular vergence
Dioptrics of deduced and reduced eyes.

Taken up within these headings are such questions as: What is an optical image and what conditions permit images to be formed? Light changes its speed in various media but why does refraction result from this fact? Why does an artificial pupil permit one to see clearly, albeit dimly, without one's usual eyeglasses? How does the concept of wavefront and wavefront curvature relate to the ability of the eye to focus? Finally, the simplifying hypothetical concepts of nodal points and principal planes are discussed and model eyeballs are introduced.

Images

Images can be thought of as collections of points such that for each point on the object, there is a corresponding point in the image—a one-to-one relationship. Although from each object point there issues an infinite number of light rays, only a fraction of these wind up represented in the image. This is so because naturally occurring light-sensitive mechanisms (eyes) collect light through apertures or pupils, thereby limiting the cone of light rays that may be received. In Fig. 2.00a, a very small pupil is shown to permit only a few light rays from any given object point to project onto the image plane. These fall at slightly differing image points. For example, point d on the object is represented in the image at d and d', as well as at every point in between. Therefore, because every point on the object is represented by an area in the image, the image is not completely sharp. However, if the aperture is small enough, the resulting fuzziness may be negligible and a relatively sharp image results, even though no lens is present. This is the principle of the pinhole camera. As the aperture increases in size, however, as shown in Fig. 2.00b, light rays from any given object point are represented throughout the image plane, with the consequence that although all points on the object are represented in the image, they are uniformly distributed, and so the image is indistinguishable from its background.

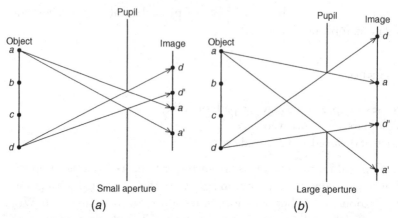

Figure 2.00. *Some Effects of Small and Large Pupils.* A small pupil such as that shown in Fig. 2.00a, permits an image to be formed. Large apertures, however, obliterate the point to point relationship required to preserve object contours in the image. With a small aperture, rays from object point *a* are spread out in the image plane between points *a* and *a'*. Likewise, object point *d* is represented by rays lying between *d* and *d'* in the image plane. Therefore, the image would not be sharp, but at least points *a* and *d* would be separated in the image plane. When pupil size is increased, as in Fig. 2.00b, the amount of image spread increases and the spatial representation of object points in the image plane no longer is discrete. For example, in the large-aperture case on the right, a ray from object point *a* is represented within the image region *d-d'*, also receiving rays from object point *d*. Thus, large apertures tend to obliterate contours by producing uniform mixing of light rays from all points on the object.

The aperture of the human eye, the pupil, thus plays a very important role as a modulator of image clarity. This may be personally attested to if during an eye examination a cyclopegic drug has been administered in order to relax the iris and open the pupil. The visual field then appears both bright and fuzzy.

Light Vergence

Points on all real objects reflect or project a diverging bundle of light rays described as having negative light vergence. When received by the human eye, the diverging bundle is limited by an aperture, controlled by a surrounding muscular, pigmented structure the iris, such that in general, the smaller the aperture, the less the bundle diverges. This is significant because diverging light must be bent or refracted in order to be focused on the retinal photoreceptors. If the negative vergence in the light bundle is too strong, it may exceed the

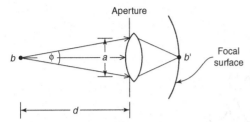

Figure 2.01. *Light Vergence and Focusing Are Reciprocal Processes.* The angle of vergence ϕ determined by the aperture size a, and distance d is a critical variable controlling the magnitude of refraction needed to focus light. The greater the angle ϕ of the diverging bundle of light from point b, the greater will be the magnitude of refraction needed to focus the bundle to a point at b'.

capability of the lens to refract the rays in order to establish clear focus. In Fig. 2.01, it may be noticed that those light rays entering the aperture near its center strike the lens perpendicular to its surface and, for reasons explained later, pass through to the imaging surface without refraction. Because looking through a small pinhole, sometimes called an *artificial pupil*, limits the incoming light rays to a small bundle that strikes the cornea approximately perpendicularly, there is much less need for refraction than there would be if the bundle were, in fact, diverged. It follows that the light rays from a given point on the object are represented very nearly in one point on the retina. As a result, targets viewed through artificial pupils tend to be seen in clear focus, even without one's usual prescription lenses in place.

In Fig. 2.01, point b on some real object reflects light rays in all directions, but only those within the angle ϕ pass through the pupil of diameter a. The angle ϕ is the measure of light vergence and is given by

$$\tan(\phi/2) = a/2d$$
$$\text{hence } \phi = 2\tan^{-1}(a/2d).$$

If the aperture a and distance d are expressed in meters, then the light vergence ϕ is given in degrees. The relationships expressed in the tangent formula among aperture size, distance, and the light-vergence angle are shown graphically in Fig. 2.02. With distance fixed, vergence varies nearly linearly with aperture size, and with aperture size fixed, nearly inversely with distance. This last relation implies that as distance to an object increases, the light vergence decreases toward zero, where the light rays remain parallel to each other.

When aperture size and vergence angle are known, distance d can be computed as

$$d = a/2\tan(\phi/2),$$

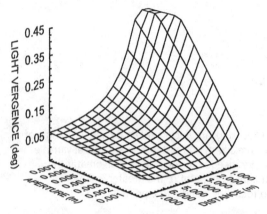

Figure 2.02. *Light Vergence as a Function of Distance and Aperture Size.*
Although negative signs are not appended to the values on the z axis, it is worth
noting that the increasing angle of light vergence refers to an increasingly
divergent light bundle, conventionally described as having negative vergence.
Because both distance d and aperture size a control the angle ϕ of light vergence,
they actually can trade off. Thus, sharp focus can be achieved by balancing a wide
aperture against a great target distance, or a near target against a small aperture.
The relationships are shown in Fig. 2.02, where the range of aperture sizes
approximates physiologic limits and the far distance (7.0 m) approaches optical
infinity. Notice that for small apertures the light vergence stays quite small and
nearly constant for the entire range of distances. Thus, the amount of focusing or
refraction needed to bring the diverging light bundle to a point on the retina also
would be relatively constant over these same distances. The range of distances over
which the required focusing power is relatively constant frequently is referred to as
depth of focus. In general, the smaller the aperture, the greater the depth of focus.

and

$$d^{-1} = 2\tan(\phi/2)/a.$$

Thus, the tangent of the vergence angle varies directly with the inverse of
distance (d^{-1}), and because for small angles the angle and its tangent are
linearly related, a linear relation also may be presumed between d^{-1} and the
vergence angle ϕ. This expression d^{-1} or $1/d$ provides a very widely used
measure of light vergence (v) whose unit is the spherical diopter (D)

$$D\,(\text{diopters}) = 1/d\,(\text{meters}).$$

The greater the light vergence, the greater the level of refraction needed to
focus light at the retina. Thus, there is good reason to develop a quantitative
measure of light vergence: Namely, to provide a measure of the work needed
to bring light rays into sharp focus.

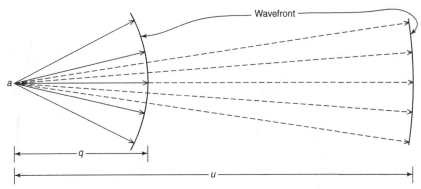

Figure 2.03. *Wavefront Curvature as a Function of Distance.* The concept of wavefront arises by analogy with transverse waves in water produced by dropping a pebble into a quiescent pond. Only a sector of the total circular wavefront is shown in the figure. The wavefronts are represented as the hypothetical line interconnecting the leading edges of light rays, all of which began traveling at the same time. Comparing wave fronts at distances q and u, it can be seen that wavefronts of the same linear extent become less curved with increasing distance.

Another more conventional way to understand the significance of the spherical diopter is in terms of the curvature of the wavefront of light. Consider in Fig. 2.03, point a from which light rays are reflected (or projected) in all directions. In the figure, only a sample of rays are represented to the right of the point after traveling a certain distance, q. Because the rays begin traveling at the same time and at the same velocity, they all travel the same distance, q. At the leading edge of the bundle of rays, represented by the arrow tips, imagine a line that interconnects each ray. This hypothetical edge, termed the *wavefront*, has a curvature that varies with distance such that the greater the distance from the source, the less is the curvature of the wavefront. This is shown visually by comparing wavefront curvature at distances q and u. It also follows formally from the definition of curvature as the rate of change in angular direction of a line, tangent to a point on a curve, moving at constant velocity through a unit length.

In Fig. 2.04, the line drawn as tangent to the curve at point p travels through a unit length s on the circumference of a circle. As it does so, it changes its direction from the vertical by an amount equal to the angle ϕ. From the calculus, curvature is defined as the rate of change in ϕ with respect to s or $d\phi/ds$.

Because, in radians

$$\phi = s/r,$$

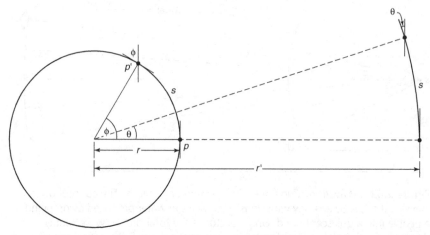

Figure 2.04. *Curvature as Rate of Change.* At the left, the tangent line at point *p* changes its angular orientation by the angle ϕ as it moves through arc length *s* to point *p'*. At the right, at distance *r'*, the tangent line changes orientation much less rapidly, as shown by the angle θ, even though it moves through the same path length *s*. The rate of change in the orientation of the tangent line defines curvature that, as the figure shows, varies inversely with the length of the radius.

then curvature,

$$K = d\phi/ds = 1/r.$$

In other words, the curvature of a circle changes at a constant rate (K) and varies inversely with the radius *r* of the circle. At a great distance, where *r* approaches infinity, the curvature approaches zero and the wavefront approximates a flat edge. We have already seen that light vergence is reciprocal to the distance from its source point, and now it has been demonstrated that wavefront curvature also is reciprocally related to the length of the radius of curvature. In fact, curvature and light vergence are isomorphic concepts

$$K \text{ (rad/unit length)} = 1/r \text{ (meters)} = D \text{ (diopters)}.$$

In general, the more diopters associated with a target, the closer is the target to the eye, and the more refraction or bending of light the eye must provide to produce a focused image of the target on the retina. In Fig. 2.05, the wavefront is drawn through the rays at distances *x*, *y*, and *z* meters, respectively, to the front surface of the cornea. At that point, the light vergence equals $1/x$, $1/y$, and $1/z$ diopters, respectively. As distance increases, it can be seen that an eye with fixed aperture receives fewer and fewer highly divergent light rays from

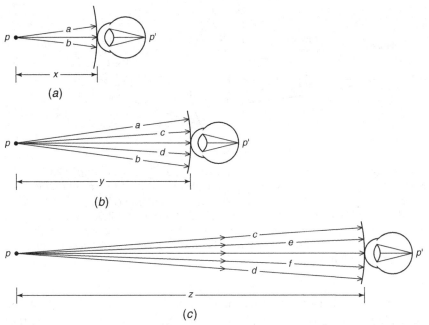

Figure 2.05. *Numbers of Impinging Light Rays Diminish with Distance.* (a) Emanating from point *p* at distance *x*, light rays *a* and *b* and all those in between are captured by the cornea, refracted toward the lens, and then again refracted to come to a focus at *p'* on the retina. (b) At distance *y*, light rays *a* and *b* miss the cornea completely. Less divergent rays *c* and *d*, and all rays in between are now focused at *p'*. (c) At distance *z*, yet fewer rays are captured by the cornea. Furthermore, because they are less divergent than at nearer distances, less refraction is demanded of the ocular lens to produce a sharp image, *p'*.

the periphery of the bundle, and therefore the curvature of the wavefront is less and less. At optical infinity, or at the distance where light vergence cannot be distinguished from that at infinity (roughly 6 meters and beyond), the angle between the limiting rays, like *e* and *f* approximate zero, they are parallel, and the wavefront is flat.

Lenses

Although a small pupil is capable of creating a relatively sharp image of an object, this approach to image formation has a serious flaw. Because small pupils produce sharp images by limiting the number of light rays being received from any one point on the object, the overall image is necessarily dim. This is so because light rays convey photons to the image plane, and therefore, the

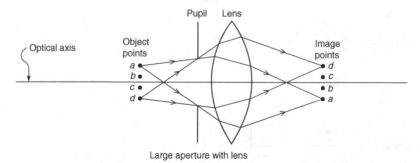

Figure 2.06. *A Positive Lens Adds Light to an Image.* A large aperture with a positive lens is capable of converging the bundle of rays from each point on the object to a discrete point in the image plane. The lens thus acts both as a collector and distributor of light rays. Without the lens, as shown in Fig. 2.00b, the large aperture would permit obliteration of the one-to-one relationship between object and image points. This would result from the mixing of light rays from different object points in the image plane.

smaller the number of light rays, the lower will be the number of photons associated with a given image point. Therefore, the sharper the image, the dimmer it becomes.

Positive lenses placed behind an aperture provide a reasonable solution to the problem of image brightness because positive lenses act as collectors of diverging light rays that become refocused as single points. They conserve the one-to-one relationship between object points and image points that is necessary for sharp and bright images. Figure 2.06 shows the limiting rays of two diverging bundles of light rays, one from object point *a* and another from object point *d*. The positive lens transforms these diverging bundles into converging ones that, in the ideal case, create a one-to-one relationship between object and image points.

Lenses, in general, are useful because they add vergence to light rays, the amount of vergence imparted being characteristic of the power of the lens. Negative lenses add negative or diverging light vergence, whereas positive lenses, like the natural lens found in the eye, add positive or converging light vergence to the bundles of light rays that impinge on it. In Fig. 2.07, light rays that emanate from a very distant point impinge with zero vergence on a positive lens. Light rays leaving the lens do so with a vergence determined by its curvature, thickness, and type of lens material. Because, in the figure parallel light rays (i.e., rays with zero vergence) are brought to a focus at a distance of 0.1 m at point *f*, the lens is said to have a power of 1.0/0.1, or 10.0 D. Such a lens adds 10.0 D of light vergence to all types and quantities of light vergence

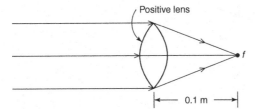

Figure 2.07. *Light Focusing by Positive Lenses.* Light rays from a point on a distant source on the left are represented by parallel lines. An ideal positive lens of 10.0*D* power focuses the light bundle at a distance of 0.1 m at *f*.

that happen to impinge on it. If a diverging bundle with a negative vergence of $-10.0D$ strikes a lens of $10.0D$ power, then the rays leaving the lens will be parallel, having a net vergence of zero diopters, as shown in Fig. 2.08. Negative lenses produce diverging bundles of light rays, just as points on real objects do. When parallel light rays are incident on a negative lens, the point from which the rays diverge defines the focal length, and the inverse of the focal length, in meters, determines the lens power in diopters. The negative lens in Fig. 2.09 has a focal length of 0.1 m and a dioptric power of $-10.0D$. Because light vergence is given by the reciprocal of the distance from the focal point, the vergence is maximal, $-10.0D$, at the lens, and decreases thereafter, as one moves toward the right of the lens.

Points on a real object at a distance of 0.1 m from the viewer present the same set of diverging bundles of light rays to the eye as those present at the locus of the lens in Fig. 2.09. In order to focus these bundles at separate points on the retina, the lens within the eye must increase its curvature and thickness by *accommodating*. Therefore, any vergence of negative magnitude (e.g., $-10.0D$) is thought of as an accommodative demand on the ocular system to neutralize the negative vergence by adding to it the same magnitude of positive light

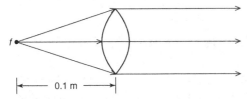

Figure 2.08. *Positive Lenses Add a Constant Amount of Light Vergence.* Diverging light rays from a point exactly one focal length from the lens leave the lens in a parallel bundle. If light emanated from a point closer than 0.1 m, the impinging light would be more divergent than shown in the figure, and the lens would no longer be capable of neutralizing it. The emerging light would then be divergent.

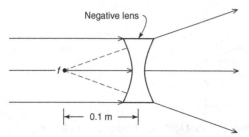

Figure 2.09. *Negative Lenses Subtract Light Vergence from the Impinging Bundle.* Light of zero vergence strikes the lens from the left; therefore, the negative light vergence emitted is due solely to the lens. The focal point *f* is arrived at by simply tracing the diverging rays backward to the point from which they *apparently* diverge.

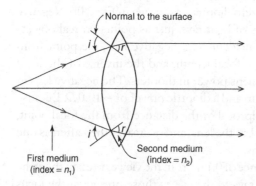

Figure 2.10. *Snell's Ingenious Law.* The Dutch mathematician Willebrord Snell (1591–1626) formulated the relationship between the angle of incidence (i) of a light ray and its angle of refraction (r) in a second medium, namely, $n_1 \sin i = n_2 \sin r$, where n_1 and n_2 are the respective indices of refraction in the respective media.

vergence (e.g., $10.0D$). That increasing lens curvature serves to increase its refracting capability is illustrated in Fig. 2.10.

Why Lenses and Prisms Change the Path of Light

The fundamental reason for the bending or refraction of light is that light changes its velocity as it traverses the media through which it moves, according to Snell's law described in Fig. 2.10. If this were not true, lenses could not collect the light rays that impinge on them and the point-to-point relation between image and object would not hold. Some totally different mechanism than we now have for seeing would be required.

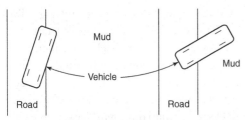

Figure 2.11. *Snell's Law for Vehicles.* The vehicle on the left enters the mud on the shoulder at a slight angle. Torque is generated by the greater traction on the left side of the vehicle, so that the car points toward the direction of a line drawn perpendicular to the highway, as represented on the right. If we knew the velocity of the vehicle, both in the mud and on the road, we could develop and test a variant of Snell's Law for vehicles to predict the precise angle of the vehicle in the mud (skidding aside).

The speed at which light traverses a transparent substance is represented by a number *n* called the *index of refraction*. It is the ratio of the velocity of light in a vacuum to its velocity in the medium of interest. The index is never less than 1 because the speed of light is greatest in a vacuum. In air, the index approximates 1 at 1.0003, whereas certain forms of glass have a refractive index of 1.5. To get an intuition of just why a change in the speed of light should result in a change in its direction, consider the physical analogy of a quantum of light energy, the photon, with a vehicle. The car, represented to the left in Fig. 2.11, has wandered onto the soft shoulder of a highway. As soon as the right front tire meets the mud surface, it sinks a bit and the forward motion of the tire meets with resistance. On the left side of the car, however, both tires remain on firm ground, retain traction and move forward at a constant rate. The net effect is equivalent to a torque or turning force with the right side being held back while the left side was propelled forward. The car then necessarily alters its initial heading to wind up, as shown on the right hand side of Fig. 2.11, turned toward the side of greater resistance. As a "gedanken" experiment, an experiment done in thought, consider the outcome of having a vehicle drive squarely straight ahead onto a mud flat, instead of entering at an angle. In this case, because both left and right tires would meet identical resistance simultaneously, no turning would result – only a slowing of forward velocity.

The purpose of analogical reasoning is to be able to apply the conclusion drawn from one domain, in this case the car, to the analog domain, in this case the photon. Therefore, one may expect that light entering a medium of greater refractive index will bend in the direction of the normal erected at the point of entrance to the second medium, whereas light striking any medium at right angles pass undeviated but at altered velocity.

The foregoing conclusions are actually true, but not perfectly logically founded. Analogical reasoning is good for providing intuitions, but not for yielding logical deductions because analogies are not themselves perfect. For example, a photon has a certain frequency of vibration at right angles to the direction in which the light is propagated. If the car in our example had bad shock absorbers, it would vibrate up and down while the car moved forward, and if the entire vehicle were to be enveloped in mud during its forward movement, a more complete analogy with a photon entering a second medium would be had. However, here we would be wrong to conclude that the vibration frequency of a photon would be slowed down in a medium of low refractive index, even though its forward motion would be retarded. The facts seem to be that the frequency of light is not changed at all as light travels from one medium to another, although its forward speed is modulated and wavelength is altered as well (Keating, 1988, p. 456). Beware of the limitations of physical analogy!

Prism Diopters and Ocular Convergence

A prism diopter (Δ) is a unit used to describe the extent to which images are shifted after refraction by a prism. Unlike lenses, prisms do not add light vergence to the impinging light bundles because their surfaces typically are flat, and therefore all light rays are bent in the same direction.

In Fig. 2.12a, the light rays are refracted twice, once on entering and again on leaving the prism. In Fig. 2.12b, the rays strike perpendicularly; therefore, refraction occurs only when the rays leave the prism on the right. If the linear shift associated either with the apparent direction of the image (shown to the left of the prism) or with the refracted ray (shown to the right of the prism) amounts to 1.0 cm for each meter of distance from the back surface of the prism, then the angular deviation α corresponds to one prism diopter. In order to capture the displaced image on the fovea, the eye must turn from the straight ahead position through the same angle α equal to 1.0^Δ. Because $\tan \alpha = 1/100$, $\alpha = \tan^{-1} 0.01$. Therefore, $\alpha = 0.573$ deg, and there are 0.573 deg per prism diopter. Note that although the light rays leaving the prism are deviated toward the base, the *apparent* location of the optically shifted target is toward the prism apex.

Because the prism diopter represents a measure of angular rotation of both an image and the eye itself, it can be used to describe the extent of binocular convergence in the sense of convergence as an angular departure of the eyes from a straight-ahead position. Figure 2.13 shows that base-out prisms

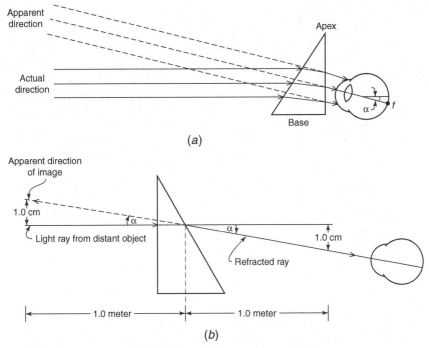

Figure 2.12. (a) *Light Bending by Prism*. Parallel light rays from a point on a distant target are all refracted by the same angle. Further refraction by the cornea and lens focuses the displaced image. The dashed lines indicate the direction in which the eye must turn to place the displaced image on the fovea (*f*). Displacement angle is represented by α. Double refraction is shown in Fig. 2.12a, and single refraction at the back surface in Fig. 2.12b. (b) *Defining the Prism Diopter*. A 1.0 diopter prism shifts the light by 1.0 cm for each meter of distance from the prism. To refixate this shifted image, the eye must rotate 0.573 deg. Note that right-angled prisms always shift the apparent direction of a target in the direction of the prism apex.

may be used to increase the magnitude of ocular vergence (viz., binocular convergence) to a particular target, and that even in the absence of prisms, the vergence angle itself may be expressed in prism diopter units.

Expressed in radians as an approximation formula, the convergence angle γ equals a/d. Thus, convergence to a given target varies directly with the size of the viewer's interocular axis a and inversely with the distance d. More precisely rendered, $\gamma = 2\tan^{-1}(a/2d)$. Solving for distance yields

$$d = a/2\tan(\gamma/2).$$

Because with γ held fixed, distance d varies directly with a, one may frame the hypothesis that people with large interocular axes see targets as somewhat

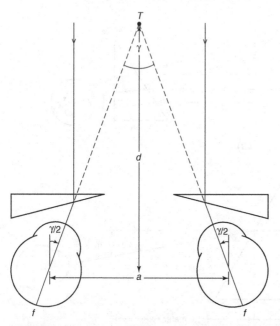

Figure 2.13. *How to Alter Ocular Vergence (Convergence).* Parallel light rays from a distant target strike two base-out prisms and deviate the light by the angle $\gamma/2$. Each eye then rotates by this amount to capture the displaced image on the fovea, f. The target T is localized at distance d, where an actual target seen without prisms would be viewed with convergence angle γ. Although ocular vergence has been altered, light vergence has not.

farther than their peers with smaller interocular axes. Is this true? Because experiments have not yet been designed to evaluate this hypothesis, we do not yet know the answer.

Dioptrics of Deduced and Reduced Eyes

In 1911, Allvar Gullstrand, a Swedish professor of ophthalmology, received the Nobel Prize in medicine for his work in physiological optics. Later, in an appendix to a translation of Helmholtz's *Treatise on Physiological Optics*, Gullstrand (1924/1962) proposed several representative model eyes that might be used for reliable calculation of the dioptrics or refraction of the human eye.

We have seen previously that the bending of light occurs at an interface where the index of refraction is different in the media on either side of the partition. In the case of the human eye, an extraordinarily large number of such interfaces exist. For example, varying indices occur between air, tear, epithelium (the transparent lining of the cornea), cornea, aqueous humor, lens, and vitreous. Furthermore, the lens has an index of refraction that progressively increases toward its center and then decreases progressively toward the posterior edge. Considering also that refraction is determined by the size of the refractive index, as well as by the curvature of the surface the light rays strike and the thickness of the medium, then one may begin to appreciate the enormity of the problem with which Gullstrand dealt.

The schematic eye represented in Fig. 2.14 was based on mensurational studies of human eyes by Gullstrand and others and also on the deductions of great mathematicians, such as Carl Gauss (1777–1855), who preceded Gullstrand in the application of their genius to geometric optics. One approach

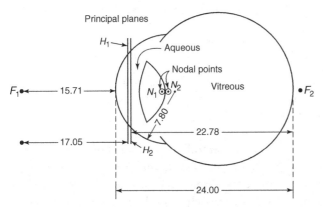

Figure 2.14. *Elements of Gullstrand's (1924/1962) Schematic Eye.* The values (mm) closely reflect the parameters of an average adult eye. Because the second focal point, F_2, is about 0.39 mm behind the retina, the schematic eye yields a slight hyperopia (farsightedness) of about $1.0D$.

Light passing through or originating from a point at F_1 diverges, but can be treated as if refracted into a parallel bundle at the first principal plane, H_1, at a distance of 17.05 mm from F_1. Therefore, the total refractive power of the model eye is 1/.01705 m, or 58.65 D. The same refractive power is present at the second principal plane, where a bundle of parallel beams may be treated as if refracted to a focus at F_2. However, because the optical distance over which light is brought to a focus varies with the index of refraction, the anterior focal distance, 17.05 mm has been weighted by 1.336, a surrogate for the several indices of refraction within the eye, to yield 22.78 mm. This is the optical distance such that, if measured in air, it would equal 17.05 mm. Accordingly, in vitreous, the power at H_2 equals 43.9 D, but in air is the same as the power at H_1.

The parameters of the schematic eye in general have been chosen so that operations performed on it, such as the calculation of image size and total refractive power, yield highly valid deductions applicable to a real eye.

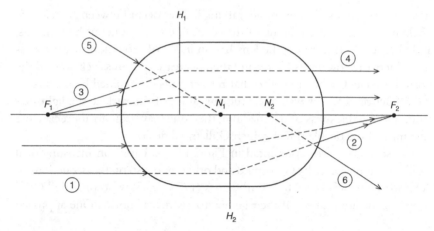

Figure 2.15. *A System of Unknown Numbers of Refracting Surfaces Separating Media of Various Indices of Refraction.* A ray such as (3) entering after passing through the primary focal point F_1 leaves the system parallel to the optical axis, as ray (4). Parallel rays such as (1) leave the system refracted and cross the optical axis at the secondary focal point, F_2, as ray (2).

Rays directed at N_1 leave the system parallel to their original path as though emanating from N_2. This characteristic defines the nodal points. Thus, rays (5) and (6) are displaced, but not refracted.

used with great success has been the method of approximation or simplification, analogous to the substitution of the cos or tan of a small angle, 1 deg or less, for its radian value, where the three agree up to five decimal places. A yet more direct example of this approach to problem solving is the concept of principal planes, an abstraction representing an infinitely thin lens without curvature but having the dioptric power equivalent to the total system it represents. These concepts are developed in Fig. 2.15.

The principal planes designated as H_1 and H_2 (from the German *hauptebene,* or *chief plane*) are the loci in space of the intersecting extensions of the entering and exiting rays. It is as if refraction occurred at these loci, when in fact, the actual bending of the rays may be quite complicated and need not be determined at all. For this reason, principal planes sometimes are referred to as *equivalent planes* because they constitute individually the refractive element equivalent to the total refractive power of the unknown system. The primary principal plane, H_1, is the hypothetical locus of refraction of rays entering after passing through F_1, whereas H_2 is the corresponding hypothetical locus of refraction for rays entering parallel to the optical axis. The concept of principal planes thus produces a remarkable reduction in complexity associated with the analysis of optical systems containing multiple refracting surfaces.

Table 2.00. *Relationships between focal points F, nodal points N, and principal planes H*[a]

$\overline{H_1F_1} = \overline{N_2F_2}$
$\overline{H_2F_2} = \overline{N_1F_1}$
$\overline{H_1H_2} = \overline{N_1N_2}$,
$\overline{H_1N_1} = \overline{H_2N_2}$

[a]Note that the bar notation refers to the distances between the designated points and planes.

Another simplifying conception is that of nodal points, hypothetical points on an optical axis such that rays directed toward the primary nodal point, N_1, behave as if emanating from N_2, the secondary nodal point. As in the case of principal planes, nodal points have no physical presence in optical systems, but the characteristics of these hypothetical structures yield a great simplification of complicated systems. It can be shown, for example, that in general, the equivalencies and relationships, shown above in Table 2.00, exist between focal points (F), principal planes (H), and nodal points (N).

In addition to the schematic eye shown in Fig. 2.15, which attempts to represent the structures and parameters of the typical adult human eye, Gullstrand and others have carried the reductionist approach forward to provide an ingenious simplification of the optics of the human eye culminating in what is conventionally labeled a *reduced eye*.

In the reduced eye (shown in Fig. 2.16), the principal planes, 0.254 mm apart in the schematic eye, are collapsed and located at a single refracting surface, the cornea. Its radius of curvature, less than that of the representative eye, has been purposely set at 5.73 mm in order to provide a total refractive power of 58.65 D, equivalent to that of the typical emmetropic eye viewing a distant target beyond 6 m (at optical infinity). Accommodation is ignored and the lens, with its continuously variable refractive index, is gone, along with all the other media, and a single refractive index, 1.336, is left in their place. As in the case of the principal planes, the small separation between the nodal points justifies collapsing them into a single locus placed at the center of curvature of the cornea. It may be recalled that light rays entering a transparent medium, perpendicular to the tangent plane at the point of entrance, pass undeviated. Because light rays directed toward the center of curvature qualify for this distinction, all so directed pass to the retina without bending, a characteristic to be expected logically if the two nodal points were sharing the same space.

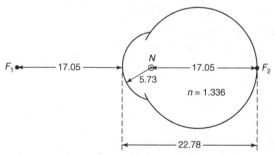

Figure 2.16. *Reduced Eye, Reflecting the Suggestions of Linksz (1950).* The reduced eye with the two principal planes collapsed at the cornea retains many of the parameters of Gullstrand's schematic eye, such as the single index of refraction, n, the distance between H_2 and F_2, as well as the focal distance between F_1 and H_1. The total power of the eye, to bring parallel light to a focus at F_2, also is approximately the same in both model eyes, namely, 58.64D given here by $(n − 1)/.00573$. All dimensions expect for the refractive index, n, are in millimeters.

This fact permits a remarkably useful simplification in the representation of the retinal location of objects and, in fact, is the logical basis for determining the apparent locus in space of stimulated points on the retina, and vice versa. All one need do to determine direction is draw a straight line from a point on an object through the nodal point to the retinal surface.

This chapter presents a snapshot of the many wonderfully complicated aspects of our image-forming mechanism and the ingenious problem-solving approaches its understanding has engendered. Next to be addressed are the remarkable oculomotor systems that have evolved to provide for functional vision in moving sentient organisms (us).

3

Oculomotor Systems

Introduction

In this chapter, the wide variety of eye movement systems is described and their functions analyzed. For discussion are such issues as:

The functional significance of moving eyes.
What enables the eye to focus.
Where the eyes converge and focus when in complete darkness.
Seeing single with two eyes.
Why some eye movements are rapid and jerky, whereas others are slow and smooth.
Eye movements that respond to linear and angular acceleration of the head.
The natural intelligence of eye-movement control systems.
Oculomotor systems in space.

Types of Eye Movements

Eye movements come with a wide variety of characteristics and seem to have evolved to provide rather specific functions. For example, *saccades*, which are flicks of the eyeball also known as rapid eye movements (REMs) when occurring during sleep, move a peripheral retinal image rapidly onto the fovea. Saccades occur when reading or whenever inspecting objects of interest (Stark & Ellis, 1981) and can be over in just a few milliseconds. Therefore, Saccades function to minimize the amount of time the eye is in flight, when retinal images may be subject to intense smear. Pursuit movements, however, permit the eye to move with the same velocity as the retinal image so that a moving target may be tracked and surveilled continuously. This holds for eye velocities as high as 100 degs sec^{-1} (Meyer, Lasher, & Robinson, 1985), but saccades can be up to ten times as fast.

Other eye-movement systems appear to make use of these slow and fast systems by combining them in a *nystagmus movement*, which is slow in one direction followed by a rapid resetting of the eye in the opposite direction when the slow component has brought the eye as far as possible in the orbit. Both the vestibulo-ocular reflex (VOR) and the optokinetic reflex (OKR) behave in this fashion, with the VOR occurring when the semicircular canals are stimulated, as when making a head movement. The OKR occurs when the whole visual field displaces while a moving observer attempts to track some object or detail in the environment. Both systems cooperate in producing gaze stability because they function to keep the gaze of a moving observer stable in space while the head is rotated or translated.

Finally, we note that because the two eyes are forward oriented in the head, it is not surprising to find eye-movement systems devoted to the management of binocular vision. The vergence and version systems thus serve to promote single vision by ensuring that both eyes point toward the same region in space. *Vergence* refers to the disjunctive movements whereby each eye moves symmetrically in opposite directions, whereas the version system controls conjugate eye movements.

The Teleology Game, or "What Is Its Function?": Two Nonmutually Exclusive Answers

What is the functional significance of eye movements? It would appear that the obvious use of an eye is to receive information about the layout of space and the identities of objects in the environment. Thus, the first answer suggests that the various types of eye movements may be understood as satisfying this information-retrieval function under certain constraints (Walls, 1942, 1962; Carpenter, 1977). For example, the information contained in a retinal image may best be transferred to the brain in a nonmoving eye, where it would be less likely to be blurred or smeared across the retina, but the eye in primates moves with the head and body. Therefore, compensatory movements that rotate the eye in opposition to head movements, or that attempt to match eye and image velocity to reduce relative motion, may have been selected through evolution because of the advantage conferred by way of a clear interpretable image. *Gaze stabilization*, the function thought to underlie these eye movements, is aided by the VOR, the OKR, and pursuit eye movement, all of which help to keep the image of an object of interest on the fovea of a moving eye. The same result also is achieved by saccades, whereby an image in the retinal periphery is brought rapidly onto the fovea. This too is consistent with the function of information extraction because parafoveal and peripheral regions of the retina, although of increased motion sensitivity, have decidedly poorer acuity than the fovea. Finally, even convergence eye movements may have been favored in evolution, both by enabling the extraction of depth differences across small intervals of space (Carpenter, 1977, p. 7) and by eliminating double images or diplopia that would occur if both eyes were not directed toward the same locus in space. There are thus numerous answers to the question of functional significance or physiologic purpose. It must be borne in mind, however, that the game of teleology (i.e., the finding of function for already existent structures) cannot be played decisively because many creative answers remain to supplant each other, and the solutions are not presently subject to empirical confirmation.

A second answer finds functional significance not only in information retrieval, but in *attribute imputation* as well. According to this attributive functional position, the oculomotor systems that are responsible for the various types of eye movements serve to impart certain attributes of space perception to the objects being regarded. For example, viewing a target while the eye undergoes smooth pursuit imparts the perception of movement to the object (Mack & Bachant, 1969), even though the target-image velocity on the retina may be near zero. Likewise, converging on a target provides a sense of distance between the observer and the fixated object (Ebenholtz & Fisher, 1982), whereas the lateral or vertical deviation of the eyes contributes a sense of radial target orientation (Ebenholtz & Shebilske, 1975; Ebenholtz, 1978; Morgan, 1978). Thus, certain attributes, such as movement or rest, movement path, distance, radial orientation, and others, are contributed, probably by way of ocular proprioception. This thesis is developed further in Chapter 4, but for the present, the point simply is that eyes both receive and impart spatial information. The reader may wish to consider the evolutionary advantages conferred by eye muscle systems that imbue objects of our world with various spatial characteristics.

Directional Eye Movements and Ocular Muscles

The eyes are capable of making movements consistent with rotations around each of the three major axes of space (i.e., vertical movements around a horizontal axis, horizontal movements around a vertical axis, and torsional movements around an axis that runs through the line of sight). Furthermore, because eye movements can be made in oblique directions (e.g., from straight ahead to upper right), they can be characterized as rotating around oblique axes as well.

Both eyes move together in the same direction in the conjugate or version mode and in opposed directions in the disjunctive or vergence mode, where the lines of sight meet at some point in front of the observer. Because, regardless of the source of innervation, the same six muscles drive the eyes, the concurrent neural messages sent by various control systems simply sum algebraically at the neuromuscular junction. Thus, a signal to converge to view a target moving toward the observer may cause the left and right eyes to deviate toward the nose, whereas a target movement toward the right may produce a simultaneous version signal just strong enough to bring the right eye straight ahead in the head. The left eye, having received both a vergence and version signal to move toward the right, would of course be deviated nasally. Such an asymmetrical posture, shown in Fig. 3.00, would appear to depart from a major rule governing eye movements – namely, Hering's law of equal innervation (Hering, 1868), whereby corresponding or *yoked* muscles of each eye are assumed to receive more or less equal innervation. Describing the law, Hering wrote:

> The movements of both eyes are to such an extent united with each other that the one will not move independently of the other; on the contrary, the musculature of both eyes reacts simultaneously from one and the same effort of the will. Accordingly, we are in general not capable of elevating or depressing one eye without the other, but both eyes are raised and lowered at the same time and in equal extent. It is equally unlikely that we can innervate the muscles of one eye alone in order to make a right or left movement. (p. 2).

Actually, both version and vergence movements separately obey Hering's law, but because of signal summation, asymmetric ocular postures may be the result (Adler, 1965, p. 411).

Each ocular globe has six extraocular muscles that, when active in various combinations, drive the eye through a certain trajectory or sustain it in a given position. Figure 3.01 illustrates the specific action of the six muscles. Note that the two oblique muscles, because they are inserted somewhat more toward the posterior of the globe equator than the others, pull the globe from the rear when contracting. As a consequence, in addition to *intorsion*, a rotation around the line of sight, top of the eye toward the nose, and an outward rotation or *abduction*, the superior oblique also causes a downward shift of the front of the

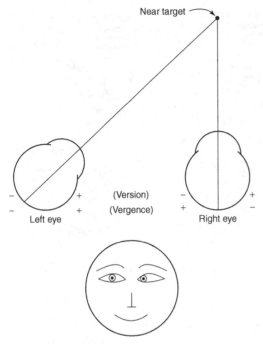

Figure 3.00. *Asymmetric Ocular Position Consistent with Hering's Law of Equal Innervation.* Convergence on the near target results from equal neural signals to the medial muscles of the two eyes, driving them in opposite directions toward the nose (vergence), simultaneously with equal neural signals to the left medial and right lateral muscles, driving the two eyes to the right (version). The plus (+) and minus (−) signs imply innervation and inhibition, respectively, on the medial and lateral extraocular muscles. In general, any binocular posture can be decomposed into a set of symmetric vergence and version movements. The alternative conception that each eye may be under separate independent control during slow asymmetrical eyemovements is a source of active contemporary research (e.g., Enright, 1996; Semmlow, Yuan, & Alvarez, 1998).

eye. In a corresponding fashion, the inferior oblique produces extorsion and an upward shift along with abduction.

It is a matter of semantics and phenomenology that the common meaning of an instruction such as "Look to the right" is associated with a movement of the *front* of the eye toward the right, whereas the rear of the eye, including, of course, the fovea, shifts toward the left. Likewise, looking up entails moving the fovea downward. Thus, we may conclude that the conventional meaning refers essentially to the position of the front of the eye, even though seeing is brought about by positioning the fovea at the back of the eye to intercept the images of objects of interest.

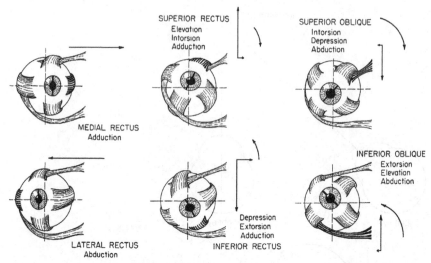

Figure 3.01. *Primary Action of Each of the Six Extraocular Muscles of the Right Eye.* (From Adler, *Physiology of the Eye,* 1965, 4th ed., p. 371. St. Louis: C.V. Mosby Co. Reprinted with permission of W.B. Saunders Co.) The medial and lateral muscles provide relatively pure horizontal rotations. The remaining four muscles each contribute a component of torsion, elevation or depression, abduction (outward or temporally), or adduction (inward or nasalward). The relative lengths of the arrows represent relative magnitudes of the three movement directions.

Muscles in the same plane of action (for example, the lateral and medial recti) obey Sherrington's Law of *reciprocal innervation* (Sherrington, 1893), whereby the agonist and antagonist are simultaneously and reciprocally innervated and inhibited. Accordingly, increasing tone in one muscle is accompanied by a proportionally decreasing tone in the opposite-acting muscle, which would otherwise prevent either smooth or rapid eye movements.

In addition to the six muscles externally attached to the globe, a seventh extraocular muscle, the *levator palpebrae,* regulates the vertical movements of the upper lid so that it elevates in conjunction with upward-directed eye movements and, on being inhibited, drops with downward directions of gaze. An eighth muscle, the *orbicularis,* under control of the seventh nerve, aids in closing the eyes. The levator and four of the six extraocular muscles are controlled by the third cranial nerve, whereas the superior oblique is directed by the fourth nerve and the lateral rectus by the sixth nerve. It is suggestive of the great functional significance of eye movements that three of the thirteen cranial nerves are devoted to their control.

Within the eye itself, a ring of pigmented tissue containing specialized dilator and constrictor muscle, the *iris,* regulates the size of the pupil, whereas a second ring of muscle tissue, the *ciliary muscle,* is responsible for changing the tension across the lens. The functional aspects of those muscle systems are now considered.

Accommodation

The Basic Mechanism in Primates

Accommodation is the process whereby the human lens changes the angle of refraction of impinging light rays. It does so by changing the curvature of the lens, mainly at the front or anterior surface. In Fig. 3.02(a), the stimulated ciliary muscle moves close to the lens, thereby relaxing pressure across the *zonules*, a set of transparent elastic tissues with relatively stiff tensile characteristics (Fisher, 1986) connecting ciliary muscle with lens capsule. This is the condition compatible with near vision, where points on an object reflect highly diverging bundles of light rays toward the eye. A suitably curved lens produces sufficient bending of light rays as to focus them sharply at the retina.

When the ciliary muscle is relaxed, as represented in Fig. 3.02(b), the muscle thins out, moves away from the lens, and tension over the zonules and across the lens capsule actually increases. The elongated lens with a large radius

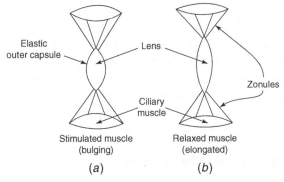

Figure 3.02. *Control of Accommodation by Ciliary Muscle.* Helmholtz (1910/1962, Vol. 1, p. 151) seems to have been the first to have worked out the main elements of the accommodative process properly, although partial insights were available several centuries earlier (Wade, 1998). When the ciliary muscle is stimulated (a), it bulges, moving closer to the lens. With tension thus lowered, the contractile lens capsule and the pliable lens cells enclosed within form a homeostatic balance, yielding greater lens curvature for near vision. With the ciliary muscle relaxed (b), it elongates, moving away from the lens and thereby stretching and reducing the curvature of the lens. This is the state of the lens when focusing on distant objects.

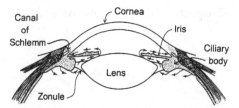

Figure 3.03. *Source and Flow Path of Aqueous Humor.* (Adapted from Davson, 1969, The intraocular fluids. In H. Davson (Ed.), *The Eye, 2nd ed., Vol. 1, Vegetative Physiology and Biochemistry*, 1969, p. 69. Reprinted with permission of Academic Press.) Aqueous humor is produced by the nonpigmented epithelium cells of the ciliary processes. Arrows indicate the flow path beneath the iris and through the canal of Schlemm.

of curvature is appropriate for the parallel or only slightly divergent light rays from distant targets because these require less bending than light rays from near targets to be focused at the retina. As you can imagine, if the cells in the corpus of the lens within the surrounding capsule were to harden, then tension on the capsule from a relaxed ciliary muscle would be to little avail in flattening the lens for far vision; nor would the inherent elasticity of the capsule be effective in near vision in increasing the lens curvature. This probably is the state of affairs in the condition of the aging eye known as *presbyopia* (see Appendix).

Finally, it is interesting to note that the same structure responsible for regulating the curvature of the lens also is the source of aqueous humor. This is produced continuously by the double layer of nonpigmented epithelium cells of the ciliary body. Figure 3.03 shows the path of aqueous humor from the ciliary body around the zonules, under the iris, and finally exiting the eye through the canal of Schlemm. A frequent cause of elevated intraocular pressure leading to glaucoma is believed to be the clogged meshwork at the entrance to Schlemm's canal.

Reflexive Accommodation and the Dark Focus

What triggers the accommodative mechanism? In reflexive mode it appears to be under negative feedback control so as to automatically minimize blur in the retinal image. When blur exceeds some threshold (Kotulak & Schor, 1986a), a neural signal is sent to the ciliary muscle to either stimulate or inhibit it, depending on what is needed to correct the focusing error. Although such error correction could take as long as 500 ms to complete, it is thought of as the rapid or phasic element in the control loop because it could not sustain the ciliary muscle tonus at the required level. By itself, it might produce clear vision momentarily, but the neural signal would decay rapidly and then the image would drift back to blur. The problem is solved with another element called a *tonic controller*, which is responsible for long-term shifts in ciliary muscle tonus. Such a slow or tonic element provides for sustained accommodation and a kind of intelligent regulation of accommodation that is self-adaptive. A typical systems diagram representing these concepts is shown in Fig. 3.04.

It is of some note that in the absence of a retinal image, as in darkness, the tonic element may be evaluated by measuring the accommodative state with no demand made on the system to reduce retinal blur. When such measurements are made (Leibowitz & Owens, 1978), the tonic controller typically is found to add about 1.5 D of accommodative power to the lens. This means that in darkness, the ciliary muscle is not fully relaxed and the lens is sufficiently curved to bring light from a target at 67 cm into sharp focus. The actual resting level or dark focus of accommodation, although quite reliable within individuals, varies considerably from person to person (Heron, Smith, & Winn, 1981; Owens & Higgins, 1983). It also exhibits a remarkable adaptive capability. For example, after about eight to ten minutes of near work, such as reading or inspecting a close target for detail, the resting level shifts inward by about 0.5 D (Ebenholtz, 1983; Schor, Johnson, & Post, 1984). Although subject to individual differences (Ebenholtz, 1985), there also is evidence that sustained fixation of a distant target produces comparable adaptation, this time shifting the resting level outward (Ebenholtz, 1983; Tan & O'Leary, 1986). Because the shifts tend to be in the direction of the previously focused targets, they have been termed *hysteresis effects* (Ebenholtz, 1983).

Shifts in tonic levels of accommodation illustrate the adaptive nature of the system because the changes serve to alleviate the workload that would otherwise be required in order to continually process errors over the negative feedback loop. The ability to adapt, especially to far targets, also may play a role in the prevention of work-induced myopia (see Appendix). Because far-target adaptation effects represent a condition of relaxed ciliary muscle, they may be especially useful in relaxing accommodation after long hours of near work. A good deal of scientific effort has been directed at determining the possible role of near-target hysteresis effects in the development of myopia, but conclusive evidence is not yet at hand (Owens & Wolf-Kelly, 1987; McBrien & Millodot, 1988; Gwiazda, Thorn, Bauer, & Held, 1991). It would be especially enlightening to know if those individuals who have escaped myopia into their adult years also fail to exhibit near-target hysteresis effects, or alternatively, if they have unusually strong far-target effects to moderate the impact of sustained near work. Other functional consequences exist for adaptive hysteresis effects in perception, and these are discussed in Chapter 4.

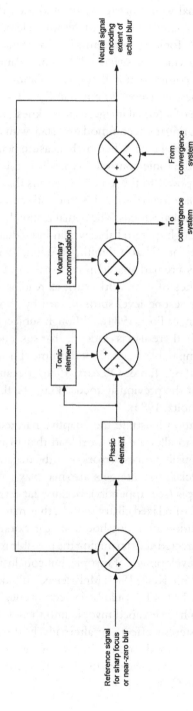

Reference signal for sharp focus or near-zero blur

Phasic element

Tonic element

Voluntary accommodation

To convergence system

From convergence system

Neural signal encoding extent of actual blur

Figure 3.04. *Accommodation as an Adaptive Control System.* A neural signal representing the actual degree of blur at some instance in time is shown on the right. This is the controlled quantity or output of the system. It also is used as feedback for comparison with a reference signal or input, representing sharp focus. The difference between the two serves as an error-correcting signal, sent forward through the phasic controller for processing mathematically (e.g., the signal may be integrated over time or simply multiplied by a constant, depending on the neural circuitry). This would determine the output in a simple negative feedback system. However, in an adaptive system, in order to reduce the need for constant comparison and correction via feedback, the output is examined and parameters are adjusted in accordance with the direction and magnitude of error correction occurring over some window in time. In the accommodation model, sampling of the error-correcting signal is done by the tonic element. If the error is nonrandom, the adaptive tonic element adds a signal to the forward loop causing an adjustment to the tonus of the ciliary muscle so as to lower the extent or magnitude of error-correcting signals that are sampled. Accommodation actually is part of a larger control system incorporating convergence because it stimulates convergence in proportion to the degree of accommodation (Flom, 1960) and also receives a boost from the convergence system to aid accommodation.

Voluntary Control of Accommodation

The ciliary muscle receives innervation from the third cranial nerve, the Edinger–Westphal nuclei in the midbrain, the ciliary ganglion, and the cerebral cortex (Jampel, 1959; Adler, 1965). It is understandable, therefore, that with special training the voluntary control of accommodation may be attained (Marg, 1951; Cornsweet & Crane, 1973; Provine & Enoch, 1975). Thus, for example, Cornsweet and Crane (1973) provided subjects with an auditory signal that was correlated with changes in the level of accommodation. This artificial biofeedback apparently was able to substitute for retinal blur, and subjects were gradually able to learn to modulate their state of accommodation in accordance with signal pitch. There is little doubt that voluntary accommodation is widespread, but no formal demographics on this point are available. Accommodation probably occurs when one voluntarily crosses one's eyes (see Fig. 3.10), a feat also relatively widespread, and in fact, there is evidence that convergence under these conditions is actually driven by the accommodation system (McLin & Schor, 1988; Ebenholtz & Citek, 1995). The arrow from the voluntary accommodation box in Fig. 3.04 is thus strategically placed right before the output to the convergence system.

It should be understood that whereas control system diagrams are in general useful for encoding large amounts of information, in the present case there are yet additional stimuli for accommodation that remain to be incorporated. One such example is change in target size that influences accommodation independently of any change in target-image blur (Ittelson & Ames, 1950; Kruger & Pola, 1985). Such a periodically looming pattern may be effective because of sensitivity to rate of change in image size, a type of optokinetic accommodative response, or the effect may be mediated by cognitive processes signalling change in relative distance between target and observer. However, the change in perceived distance may itself be determined by the change in accommodation occasioned by the looming pattern. It may be worthwhile to be sensitive to these "chicken vs. egg" possibilities when attempting to identify causal relationships between physiologic mechanisms and perceptual states.

Pupillary Movements and the Iris

The *iris* is the colored portion of the eye, the open center of which is the *pupil*. At birth, eyes of all people are various shades of blue simply because long wavelengths of light (in the red region) are differentially absorbed before the remaining wavelengths are reflected back from the iris. Soon after birth, pigmented cells, *chromatophores*, develop on the rear surface of the iris, the portion that floats just above the lens. The density of chromatophores determines the color of the adult eye (Adler, 1965).

Two types of muscle, a *dilator* and *constrictor* or *sphincter*, control the quantity of light in the retinal image by varying the pupil diameter from about 1.0 to 8.0 mm. Because the retinal image is at a fixed distance from the pupil, of about 20 mm (see Chapter 2), the retinal intensity varies with the area of the aperture and hence with the square of the aperture radius. This provides a range of about 64:1 in retinal illumination. In terms of a camera analog, because f-number is defined as the ratio of focal distance to aperture diameter, the pupil provides an f-stop range from about f/20 to f/2.5, the latter being not quite as bright as the value of f/1.9 available in many cameras.

Functional Aspects

The pupillary light reflex has obvious functionality in that it brightens an image seen under dark illumination and reduces image brightness under high illumination conditions. This simple view is somewhat complicated, however, by the fact that the reflex is modulated by light adaptation, becoming less responsive as the retinal receptors decrease in light sensitivity (Lowenstein & Loewenfeld, 1959). Generally speaking, as age increases beyond about 18 years, pupillary responses diminish, yielding typically smaller pupils in dim illumination.

Many conditions influence pupil size, including fear, anxiety, pain, and vestibular and other sensory stimulation, all of which induce pupillary reflex dilatation. However, the pupil is reduced in size during sleep and in narcosis (Lowenstein & Loewenfeld, 1969). The neural transmitter acetylcholine, associated with the parasympathetic branch of the autonomic nervous system, generally is responsible for pupil constriction, whereas the sympathetic transmitter adrenalin facilitates dilation. It is easy to see, therefore, why a great variety of emotional states influence pupil size.

The pupil also takes part in what has been termed the *near reflex*, a three-part synergism between accommodation, vergence, and pupillary reflex. Thus, when viewing a nearby target, the two eyes converge, each eye accommodates, and the pupils constrict. As shown in Chapter 2, although a small pupil reduces retinal illuminance, it also has a very positive functional consequence – namely, that of reduction of image blur. Put another way, a small aperture increases the depth of focus, thereby obviating the need to clarify an otherwise blurred image by additional focusing.

Vergence

Definition, Functionality, Decussation, and Stereopsis

Vergence eye movements are disjunctive in the sense that each eye moves in a direction opposite that of the other eye (Fig. 3.05). They are primarily associated with species having forward-oriented eyes, such as primates and members of the cat family (Hughes, 1972), and probably represent a working solution to the problem of how single vision emerges from two eyes and therefore from two images of the same object (Walls, 1962). Because vergence movements ensure that for the object of interest the image in each eye falls on corresponding loci of each retina, both of which project to the same region of brain (Figs. 3.06 and 3.07), sensory fusion of the images ensues. In those species with panoramic vision, like the rabbit, where the eyes are almost 180 deg apart, totally different images are represented in the brain from objects within each visual field, but because each eye receives images from different objects, the problem of *diplopia* (double images of the same object seen at different locations) would not exist. However, many lateral-eyed species, including the rabbit, have a region of binocular vision where each monocular field overlaps the other to some extent. In this case, an object within the binocular field stimulates both eyes, but instead of a vergence eye movement being initiated, the whole animal may jump or, as in the case of the lizard or frog, a tongue may strike out to capture the object of interest.

It also is distinctly possible even in lateral-eyed species that a binocular field, where a single object is represented in images of both eyes simultaneously, forms the basis for *stereopsis*, the visual appreciation of solidity or of a depth interval between objects (Hughes, 1977, p. 642). The minimum requirement for stereopsis is that noncorresponding points be stimulated so as to produce a slight binocular disparity. In humans, this implies stimulation from a point on the horopter (see Fig. 3.06) and another at some small distance within or beyond it to create the disparity. Of course, if these disparate points are separated too far, they produce double images; however, if within a threshold region, the disparate points fuse but also signal a sense of depth. Overlapping visual fields, usually but not universally associated with vergence, may be a precondition for binocular disparity and depth. A quantitative measure of binocular disparity is derived in Fig. 3.08.

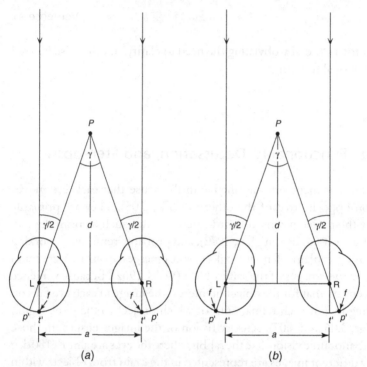

Figure 3.05. *Approximate Geometry of Vergence.* (a) The visual axes are parallel, the left (L) and right (R) eyes aimed at a distant target *T* (not shown) whose images (*t′*) fall on the fovea *f*. A nearby object *P* at distance *d* results in disparate images at *p′* represented to the left of the fovea of the left eye and to the right of the fovea of the right eye. If the two images are perceived, the condition is labeled *diplopia*, or double vision. After a reaction time of 160 ms (Rashbass & Westheimer, 1961), a decision to look at object *P* (b) causes reflexive rotations of each eye that serve to place the images of *P* on each fovea. Because *P* is then seen as a single object, the movements have been described as a *fusion reflex*. Note, however, that *T* is now represented in disparate images. These disparate images constitute a major stimulus for vergence eye movements (Westheimer & Mitchell, 1956, 1969), but beyond about 4 deg of arc, disparity diminishes in effectiveness (Erkelens, 1987) as a stimulus to vergence movements. Assuming an axis of rotation about 13.5 mm behind the cornea (Alpern, 1969) and given the separation between the axes *a*, the angle of convergence *γ* may be estimated in degrees as:

$$\gamma = 2\ \tan^{-1}\left(\frac{a}{2d}\right).$$

Alternatively, assuming *d* to be a radius of curvature and *a* therefore a section of arc, then:

$$\frac{\gamma}{360°} = \frac{a}{2\pi d},$$

therefore

$$\gamma = \frac{a}{d} \times 57.3.$$

Horizontal vergence is thus defined in terms of the angular departure of the visual axes from the straight-ahead position.

Vergence-Stimulated Accommodation

When vergence movements occur they also provide an assist to the accommodative mechanism. With increasing vergence, accommodation also increases so as to be appropriate for near vision. Likewise, with decreasing convergence, accommodation relaxes for distant vision. This means of altering accommodation seems redundant, given the negative feedback control that focusing already has (see Fig. 3.04). However, there is added functionality here in that this convergence signal to accommodation is a *feed-forward* signal. Without the feedback portion of the loop, feed-forward signals tend to be much faster than those of the feedback variety. Therefore, the focusing mechanism is made ready for changes in distance without first having to sample the blur in the image and then send through an error-correcting signal.

Individuals differ in the quantity of accommodation triggered by convergence. A low ratio of convergence accommodation per unit of convergence (CA/C ratio) requires more effort from the negative feedback loops of the focusing mechanism than would a high CA/C ratio. This added burden is a potential cause of *asthenopia*, or eyestrain (see Appendix). It is interesting that adaptive systems characteristic of oculomotor control seem to be organized around the general principle that wherever possible, feedback control should be kept to a minimum. Frequently, this is accomplished by substituting altered values of feed-forward signals in their place so that feedback loops may become quiescent. The stimulation of error-correcting feedback loops may thereby serve as the precondition for subsequent adaptation, whereas altered values of parameters in the feed-forward portion of the control loop constitute the adaptive response itself.

As a case in point, consider that the CA/C ratio itself is subject to adaptation. Miles, Judge, and Optican (1987) had subjects wear a device, represented in principle in Fig. 3.09, that effectively increased the interocular axis (the separation between the eyes). Thus, the device increased, by a multiplicative factor the convergence angle of every object the subject viewed. Because of the convergence-accommodation link, the device increased the accommodation as well. Such an unnecessary increase in accommodation, along with error-correcting feedback, probably triggered an adaptive response in the focusing system requiring less accommodation for each unit of convergence than previously. Lowering the level of accommodation would then yield a lowered CA/C ratio, which in fact was found.

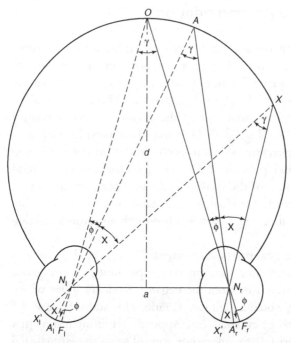

Figure 3.06. *The Horizontal Horopter.* (From Linksz, 1952, *Physiology of the Eye,*
Vol. 2, p. 441. New York: Grune & Stratton. Reprinted with permission of W.B.
Saunders Co.) Within an error of about 7.8 mm it is useful to assume that the
schematic nodal point, 5.7 mm behind the cornea, and the center of ocular
rotation, 13.5 mm behind the cornea, exist at the same point in space. Given this
fiction, then an equi-convergence curve can be drawn through the nodal points N
having the simple shape of a circle. Binocular fixation on any point on this circular
locus produces the identical convergence angle, as does fixation at any other point,
as long as the distance d is not altered. Another significant property of this curve is
that the angular separation between images of points in one eye are exactly equal
to their separation in the other eye. For example, the angles ϕ in the two eyes must
be equal because they are included angles intercepting the same arc, OA. Because
point O is at the fovea F and the images of O and A are equidistant in each eye,
they can be identified as falling on corresponding retinal points and therefore are
seen as fused or single. Points in space, the locus of which forms a circle passing
through the nodal points and stimulate corresponding retinal points, define a
boundary or horopter of single vision. Although the term was coined and the
concept first introduced by a Jesuit monk, Francisco Aguilonius, in 1613 (von
Noorden, 1980, p. 21), the *horopter* was first properly described by Johannes
Müller (1826) and G. U. A. Vieth in 1818. Accordingly, this form of the horopter
has become known as the Vieth–Müller circle. It is interesting to note that
stereoscopic depth perception results when points are present at some small
distance in front of or behind the horopter locus, whereas large displacements
produce noticeable double images.

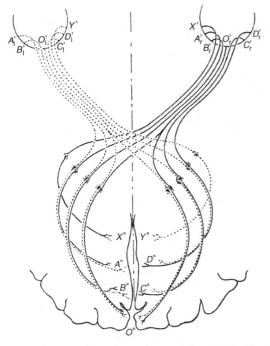

Figure 3.07. *Decussation.* (From Linksz, 1952, *Physiology of the Eye, Vol. 2,* p. 350. New York: Grune & Stratton. Reprinted by permission of W.B. Saunders Co.) *Decussation* refers to the crossing of lines as in the form of the Roman X. In birds and fish the optic nerves completely cross to opposite halves of the brain. In primates there is a partial decussation, in which, according to various accounts (Blinkov & Glezer, 1968), about 40 percent of the optic nerve fibers stay on the same side as the eye from which impulses are transmitted, whereas the remaining 60 percent cross over. According to this arrangement, properly described by Newton (1730/1952), the left half of *each* retina transmits to the left cortex and the right half is linked with the right occipital lobe. Points such as *A, B, O, C,* and *D* on each retina, that when stimulated, signal to nearly identical brain regions are known as *corresponding points.* An object, the image of which stimulates corresponding retinal points, tends to be seen in the same direction in space from each eye, and, because of sensory fusion, also is seen in single vision. The site at which the projecting fibers synapse actually is a six-layered relay station, the *lateral geniculate body.* Fibers from the ipsilateral eye populate layers 2, 3, and 5, whereas the contralateral hemiretinas are represented in layers 1, 4, and 6.

Voluntary Vergence

Disparity-driven vergence is essentially reflexive in that it occurs without conscious supervision, but it is possible to add a volitional element of control. Thus, even in complete darkness, many individuals may converge and relax at

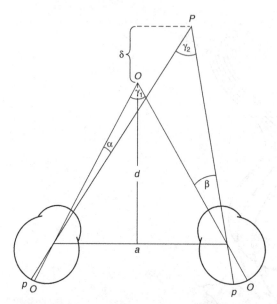

Figure 3.08. *Binocular Disparity.* Because coordinated vergence movements keep the two foveas closely oriented on the same detail *O*, other details such as *P*, in a different depth plane not on the horopter, stimulate noncorresponding points, thus causing binocular disparity and stereopsis. The difference in retinal angles ($\beta - \alpha$) defines the magnitude of disparity, traditionally designated (η). An expression for η may be derived from the identity ($\beta - \alpha$) = ($\gamma_1 - \gamma_2$). As shown in Fig. 3.05, in radian measure:

$$\gamma_1 = \frac{a}{d},$$

and therefore as an approximation

$$\gamma_2 = \frac{a}{d + \delta}.$$

Therefore,

$$\gamma_1 - \gamma_2 = \frac{(ad + a\delta - ad)}{d(d + \delta)} = \frac{a\delta}{(d^2 + d\delta)}$$

Assuming δ to be small relative to d allows one to drop the $d\delta$ term, yielding:

$$\eta = \frac{a\delta}{d^2}.$$

Thus, disparity varies inversely with the square of the distance to the nearer target and varies directly with the length of the interocular axis *a*, and the actual depth interval δ. The equation also indicates that those individuals with the larger interocular axis *a* will have the greater disparity η. Because disparity controls the amount of depth experienced, one might expect adults to experience a greater sense of depth than young children, whose interocular axes are relatively small, but this has not yet been proved to be the case.

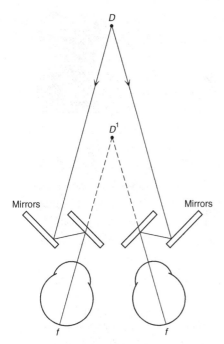

Figure 3.09. *Stretching the Interocular Axis.* (Adapted from Fisher & Ciuffreda, 1990, Adaptation to optically increased interocular separation under naturalistic viewing conditions, *Perception, 19,* 171–180. Reprinted with permission of Pion, London.) The interocular axis may effectively be increased by presenting to each eye the view it would receive if the eyes were more separated than they actually are. In the figure, the outer mirrors are separated by about twice the typical interocular separation of ~63.0 mm. These images are then reflected into the eyes at fovea *f* via the pair of inner mirrors. Because convergence varies directly with the size of the interocular axis (see Fig. 3.05), the target at *D* would be seen with twice the convergence it would elicit if seen directly without mirrors. This is equivalent to viewing the target as if it were at *D'* one-half the distance to *D*.

will and, of course, repeat the feat in light as well. Figure 3.10(a) shows the author's eyes verging and Fig. 3.10(b) shows the eyes looking straight ahead in darkness in pictures taken under infrared illumination. Voluntary vergence can be trained and these procedures are helpful in the treatment of vergence disorders such as vergence insufficiency (Cooper et al., 1983). Until recently, it was not entirely clear, however, whether such voluntary control is exerted over the vergence system directly or indirectly via accommodation because both responses tend to occur together. McLin and Schor (1988) found evidence for an indirect route when they compared the ratio of convergence to accommodation measured during the voluntary effort in darkness to the ratio

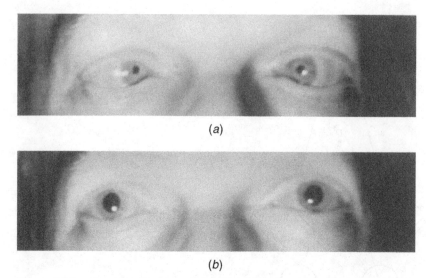

(a)

(b)

Figure 3.10. *Voluntary Vergence in Darkness.* This infra-red photo shows (a) that a pure voluntary vergence can be sustained for at least 8 min in complete darkness (Ebenholtz & Citek, 1992) and, as the small pupil suggests, that the pupillary diameter correlates highly and negatively with changes in voluntary vergence in darkness. Both the pupil size and vergence probably are driven by voluntary accommodation (McLin & Schor, 1998). (b) Note the large pupils on instructions to look straight ahead. After about 2 min of sustained vergence in darkness, subjects loose awareness of where their eyes are pointing; nevertheless, vergence is maintained. Thus, vergence control goes on independently of vergence-posture awareness. Contrary to mothers' warnings, the eyes did not remain locked in the verged position when the session was over.

obtained in response to a blur stimulus. The two ratios were highly correlated ($r = 0.9298$) and not significantly different from each other, whereas the ratio of accommodation to convergence obtained in darkness and that measured in response to binocular disparity were poorly correlated and significantly different. Similar results were found for rhesus monkeys trained to verge voluntarily in darkness to a remembered target location (Gnadt, 1992). Therefore, it appears that when one voluntarily converges, the process is mediated by voluntary accommodation (viz., it actually is voluntary accommodative vergence).

Vergence Resting Level and Its Adaptation

Where do the eyes verge when no object is present to produce disparate stimulation? If one is under deep anesthesia, comatose, recently dead, or in deep sleep, a divergent position is likely (Adler, 1965, p. 490). When one is reasonably alert, however, the matter is quite different. When subjects are placed in darkness and instructed to look into the distance (Fincham, 1962) or to relax (Owens & Leibowitz, 1980), an intermediate vergence posture typically is assumed. For example, Owens and Leibowitz (1980) found a vergence distance of 116 cm averaged over 60 subjects. For a few subjects, the lines of sight were actually parallel, as if regarding a target at optical infinity, but for most the eyes converged at about 1.0 m, somewhat beyond arm's length. Thus, there are strong individual differences in the vergence resting level, but there also is some evidence for its reliability because individuals retain their relative ranking when compared over hours as well as weeks (Fisher, Ciuffreda, Tannen, & Super, 1988).

The resting vergence as one of four reliable parameters of the visual system is represented in Fig. 3.11. In the figure, the numbers used for resting vergence and resting accommodation are illustrative only, because actual demographics on large populations are not presently available, but they do approximate available data (Owens & Leibowitz, 1980). What does the vergence resting level do? Current understanding is that like the resting focus (Ebenholtz, 1992), resting vergence serves as a zero reference level or set point with respect to which convergence effort is modulated. Because vergence effort is correlated with eyestrain or asthenopia (Owens & Wolf-Kelly, 1987; Tyrrell & Leibowitz, 1990), it is expected that sustained vergence on a target is more likely to produce symptoms of visual fatigue the greater the optical distance between the target and the vergence resting level. For this reason, operators of video display terminals may wish to place the screen at their individual vergence resting level to avoid the symptoms of eyestrain.

In general, whenever the capability of disparate stimuli to drive vergence is challenged, the vergence posture moves toward its default position, the resting level. This occurs under a variety of conditions, such as alcohol intoxication, hypoxia (oxygen depletion), dim illumination, or barbiturate intoxication (Owens & Leibowitz, 1983). Because the resting level itself tends not to be influenced by alcohol (Miller, Pigion, & Takahama, 1986), and probably not by these other conditions as well, objects represented beyond threshold distances, either nearer or farther than the resting level, produce double images, an obvious hazard to personal navigation.

It is significant that when vergence is sustained at target distances other than that of the resting level, an adaptive response occurs – a fusional aftereffect that causes the resting level to shift toward the target, thereby requiring less effort to keep the eyes converged at the working distance. It is as if there were an aftereffect on the extraocular muscle tonus, an adjustment to the muscle

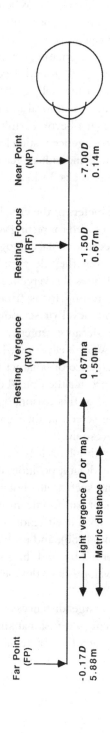

Far Point
(FP)

Resting Vergence
(RV)

Resting Focus
(RF)

Near Point
(NP)

-0.17*D*
5.88m

0.67ma
1.50m

-1.50*D*
0.67m

-7.00*D*
0.14m

— Light vergence (*D* or ma) →
— Metric distance →

D: diopters
ma: meter angles
m: meters

Figure 3.11. *Four Visual Parameters in a Slightly Near-Sighted Eye.* (From Ebenholtz, 1991, The effects of teleoperator-system displays on human oculomotor systems, SAE paper number 911391, copyright 1991, Society of Automotive Engineers, Inc. Reprinted with permission from SAE.) The nearest and farthest distances at which a target produces minimal retinal blur are the near and far points, respectively. Far point is 0.17*D* short of optical infinity (0.0*D*); therefore, the eye is slightly myopic. Resting focus (RF) derives from the tonus of the ciliary muscle (see Fig. 3.02) when no target is present, in darkness. If a target were present at the distance indicated (0.67 m), it would be sharply focused. Resting vergence (RV), also described as the *physiologic position of rest* (Vaegan, 1976) and as *tonic vergence* (Maddox, 1893), is the vergence posture adopted in the absence of any fusional stimuli, without a stimulus to accommodation, and with no voluntary effort to converge. It is interesting that RV and RF do not coincide but typically separate, with RF closer than RV. Although only weakly correlated (Owens & Leibowitz, 1980), they *do*, however, share a common noise level in that both oscillate at about 0.19 to 0.38 Hz (Kotulak & Schor, 1986b).

innervation that automatically sets the tonus to approximate the requirements for the sustained convergence distance. A variety of visual tasks are sufficient to trigger a fusional aftereffect, a phenomenon known since the founding of the discipline of physiological optics (e.g., Hering, 1868, p. 34; Maddox, 1893; Bielschowsky, 1938.)

A critical feature of the aftereffect is that after even a few minutes of sustained single vision, it does not wear off when the fusional stimulus is removed. For the same reason that it becomes less effortful to stay converged on target, it is more effortful to verge on a new target. In fact, frequently after a long period of reading or other intense near work, looking away into the distance produces diplopia for some time, even in the presence of other fusional stimuli in the distance. In this case, because vergence is pulled in the direction of the previous resting level, the eyes converge in front of the distant target, nearer to the observer, a condition known as *eso fixation disparity* (Fig. 3.12). If one eye is then closed or the observer goes to sleep, thereby preventing other fusional stimuli from acting to reset the vergence resting level, the condition may last for hours (Carter, 1963). This adaptive process is thought to be quite significant for maintaining the coordination of binocular vision so as to avoid diplopia during sustained near or distant work. Vergence adaptation thus permits automatic adjustments to be made for an otherwise inappropriate binocular balance. Unfortunately, for reasons not presently understood, not all individuals are capable of adapting their resting or tonic vergence. For these individuals, continuous error correction due to negative feedback in the disparity vergence control system occurs without the benefit of adaptive tonic vergence to reduce the feedback effort. Asthenopia (see Appendix) is a likely consequence (Ebenholtz, 1988).

The Saccadic System

Saccadic Suppression

Before describing the saccadic mechanism itself, it is worth noting a curious phenomenon associated with these rapid eye movements. When a saccade is made in the presence of illuminated objects, the images of these objects move on the retinal surface as rapidly as the eye movement itself. Yet it requires special circumstances, such as a bright luminous target in a dark surround, to perceive the smeared, stretched out images. Imagine the distraction that would accompany each attempt to look around one's environment should these images become apparent. It is therefore of obvious functional significance

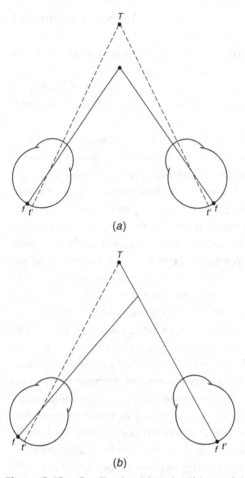

(a)

(b)

Figure 3.12. *Eso Fixation Disparity.* This condition may be induced after a short period of sustained convergence at a near target. Measurements in darkness, by infra-red photos or other techniques, would show the eyes verging to a nearer point than they did before the sustained vergence period. This represents the inward (eso) shift in vergence resting level. With a distant target present, however, the vergence aftereffect would be manifest as a fixation disparity caused by a distant fusional stimulus unable to compete fully with a strong near, tonic vergence posture. (a) Both eyes are shown verging symmetrically in front of distant target *T*. Because *T* is represented on neither fovea (*f*), a fixation reflex would be stimulated so as to place the image of *T* (*t'*) on the fovea of the dominant eye. Assuming this to be the right eye, fixation may be accomplished by a version movement of both eyes to the right, as shown in (b). Thus, what is basically a symmetric shift of both eyes toward the resting vergence position comes to have the appearance of an asymmetric ocular misalignment. With continued exposure to the target, the fixation disparity will gradually diminishes (Schor, 1979) and a new vergence resting level emerges. Dashed lines represent the path of light rays from *T* to the retina. Ocular directions are represented by solid lines running from fovea through the nodal point to the target.

that a mechanism exists to inhibit the detection of the image path caused by a saccadic eyemovement. This incompletely understood phenomenon, *saccadic suppression*, refers to the fact that our ability to detect the presence of a target is severely reduced, not only during a saccade, but also up to about 75 ms before the saccade actually begins and for a somewhat greater time interval after it ends, with intensity thresholds elevated by more than 0.5 log units (Matin, 1986). Although an image moving within the range of speeds for saccades but with a stable eye may also contribute to such inhibitory effects via metacontrast masking (MacKay, 1970; Matin, Clymer, & Matin, 1972), other carefully controlled studies that rule out masking suggest a central inhibitory factor as well. For example, Riggs, Merton, and Morton (1974) demonstrated suppression of electrically stimulated phosphenes, sensations of light originating entoptically (within the eye), produced during saccades in total darkness. It seems reasonable to conclude that without these inhibitory processes, the very functional nature of saccades, to move images to the fovea for information processing, would be severely disturbed. Saccadic suppression thus remains an obviously functional mechanism, in need of further explication.

Pulses and Steps

In neurologic terms, a *pulse* represents a large set of neurons firing asynchronously at a high frequency over a certain time interval. When the agonist extraocular muscles receive such a pulse, a corresponding set of muscle fibers contract in proportion to the firing rate and the total number of neural impulses received, whereas neural stimulation to the antagonist muscles is temporarily halted (Leigh & Zee, 1984). The longer the duration of the pulse or burst, the greater will be the amplitude of eye turn, whereas eye velocity is controlled by the firing rate and number of motoneurones recruited (Bahill & Stark, 1979). As a result of this arrangement, extremely rapid changes in the position of the eyeball in the orbit may occur. However, if not for a second neural signal, the *step*, elastic restoring forces would return the eye to an equilibrium position, much like a rubber band after release of tension but considerably more slowly. This status is prevented by new steady-state tonic levels in both agonist and antagonist muscles serving to maintain the eye in its new position when the pulse is terminated. Saccades thus consist of combinations of phasic and tonic states driven by neural pulse and step signals, respectively. These are represented in Fig. 3.13.

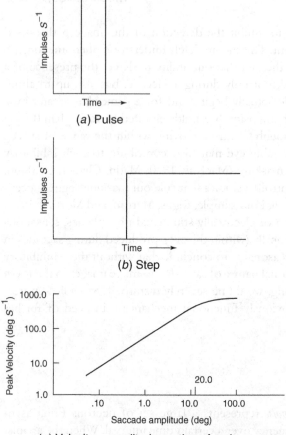

(a) Pulse

(b) Step

20.0

(c) Velocity – amplitude portion of main sequence

Figure 3.13. *Neurologic Pulse and Step Signals.* Specialized burst neurons produce a pulse of activity in oculomotor nerves (a) that lasts usually for about half the time the eye is in motion (Bahil, Clark, & Stark, 1975). Slightly before the end of the pulse signal, step signals (b) to both agonist and antagonist eye muscles help lock the eye in the position to which it was brought by the pulse. The greater the duration of the pulse, the greater will be the amplitude of eye movement. Furthermore, both the firing rate and the numbers of motoneurones recruited also increase with pulse duration. Therefore, both duration of the saccade and its peak velocity are monotonically related to saccade amplitude. Bahil et al. (1975) named the plot of these relationships the *Main Sequence* (c) by analogy with astronomers' use of the same term to represent the relationship between star brightness and color temperature.

When a pulse and step are mismatched, so that the position to which the eye is brought by the pulse is different from that entailed by the step, slow drift movements (glissades) occur to bring the eye toward the ocular position determined by the step signals. Some of these dysmetric saccades are indicative of brain dysfunction and play a role in diagnosis. Likewise, saccades that deviate from the main sequence may signal dysfunctional brain states.

Basic Attributes

These very rapid eye movements can have a velocity as high as 900 deg sec^{-1}. Most naturally occurring saccades, including those associated with reading, fall in the range from 4 min of arc to about 15 deg (Bahill et al., 1975), the latter taking about 50 ms to be completed. The underlying neural control system is remarkable for the exquisite timing relations required between agonist and antagonist eye muscles to initiate the movement and then to bring it to a rapid stop at or near its target. However, we describe in Chapter 5 the possibility that a dyschronicity in stimulation and relaxation of opposed eye muscle groups may actually lead to motion sickness.

As noted previously, one of the functional aspects of saccades is to place an image represented in the retinal periphery onto the central fovea for clear viewing. It typically takes about 200 ms for brain circuits to calculate the direction and magnitude of the required saccade, but modifications to the instruction set used to create the saccade can be made up to about the last 80 ms prior to the saccade onset (Becker, 1991, p. 128). For example, imagine an eye fixating straight ahead when a target light comes on at 30 deg to the right. Normally, the saccade would be initiated after about a 200 ms "countdown." However, if a second target light now at 15 deg to the right is shown, the saccade to the first target may be modified, depending on how far into the countdown the second target was triggered. With only 80 ms left, the second target would be without effect, whereas with nearly the full 200 ms available the saccade would be to the second target, ignoring the first. With intermediate values of this "reprocessing time" (Becker, 1991, p. 128), a saccadic transition zone would be traced out with the eye saccading to intermediate positions. Another instance of the remarkable computational apparatus underlying saccades is afforded by the example of a saccade whose initial movement is used to turn off the target to which it is directed. These saccades made in darkness typically lead to corrective saccades that actually increase the accuracy of eye position relative to the hidden target. These results suggest the presence of a proprioceptive or extraretinal feedback loop that along with visual stimuli guides the saccadic computations (Becker, 1976; Shebilske, 1976). A similar conclusion is supported by the result that a brief flash on the fovea during a saccade to a previously extinguished target produced an accurate secondary saccade (Hallett & Lightstone, 1976). After the primary saccade, a secondary saccade was made back to the position in space from which the second flash emanated. Because no retinal error existed to signal a direction and magnitude, computations for such an accurate eye movement would seem to require information about the eye position in the orbit when the second target was flashed. Despite such evidence for proprioceptive control of saccadic eye movements, it is a separate question as to whether proprioceptive or extraretinal eye movement signals are available for perceptual judgments, such as those involved in the perception of radial direction. These and related issues are discussed in Chapter 4.

Are all saccades alike? It is worth noting that saccades that occur in a wide variety of circumstances do not all have the same velocity characteristics. Saccades targeted to visual objects tend to have the highest velocities, whereas saccades occurring in the absence of a visual target, under voluntary control (Henriksson, Pyykko, Schalen, & Wennmo, 1980), when the target is represented in memory (Sharp, Trovst, Dell'Osso, & Daroff, 1975), or when the saccade occurs as the fast phase of vestibular nystagmus (Ron, Robinson, & Skavenski, 1972) tend to be considerably slower. We may infer that the shaping of the duration of the neural pulse thus reflects the presence or absence of the visual target.

Adaptation within the Saccadic System

One of the earliest laboratory demonstrations of adaptive saccades was by McLaughlin (McLaughlin, 1967; McLaughlin, Kelly, Anderson, & Wenz, 1968). After training to saccade between two targets 10 deg apart, a third target was illuminated in different studies at a distance of only 5 deg or 9 deg from the initial fixation point, whereas the second target was switched off. Because the new target came on right after the start of the saccade, subjects at first made overshoots to the location of the old target with the subsequent corrective movements back to the new target. Within about ten to fifteen trials, however, saccades became tailored to the new distance, moving toward a typical hypometric or shortened profile (Henson, 1978). These results suggest that at least the gain of the pulse phase of the saccade is adaptable. In other words, the ratio of the initial saccade amplitude to the retinal separation of the points between which the eye saccades is modifiable. Similar plasticity has been observed in patients with partial abducens (sixth) nerve paralysis, in which weakened horizontal eye muscles would at first produce severe undershooting of the target in the paretic eye. With the nonparetic eye covered, the saccadic system adapted according to the needs of the weakened eye so that saccades became accurate. Evidence for changed system parameters became apparent, however, when the paretic eye was patched. As expected from Hering's law of equal innervation, the nonparetic eye then showed hypermetric characteristics (overshooting) followed by drift toward the target (Kommerell, Olivier, & Theopold, 1976; Leigh & Zee, 1984).

Once again, eye movement systems illustrate remarkable adaptive capacity to overcome error. In principle, the source of the eye movement error, whether brain lesion or muscle disorder, is transparent to the adaptive process unless the lesion occurs in brain circuits used for the adaptation process itself (Optican & Robinson, 1980). Irreversible malfunction then ensues.

The Optokinetic Reflex (OKR), Smooth Pursuit (SP), and the Vestibulo-Ocular Reflex (VOR)

Functional Significance

An object of interest near an observer in a predatory environment must be identified, perhaps as something to eat or as something to escape from so as to avoid being eaten. Because image movement degrades acuity, to facilitate inspection, even in a benign environment, it would be of great utility to have the image of this object remain stable with respect to the retina. Three inter-related oculomotor systems, smooth pursuit (SP), optokinetic reflex (OKR), and vestibulo-ocular reflex (VOR), contribute to this goal. Two of them, SP and OKR, achieve retinal stability by matching the eye velocity with that of the image through the use of neural mechanisms that sense the difference in velocity so as to reduce this difference. The third system, the VOR, functions for nonmoving targets fixated by a moving observer. The VOR fixes the eye in space by permitting the skull to rotate around the eyes, thereby promoting gaze stability and visual fixation.

Optokinetic Reflex

In the optokinetic (OKR) case, when a large portion of the retina is covered by the moving image, as when the head turns from left to right in a structured environment, an optokinetic stimulus may be identified that automatically triggers a slow eye movement in the same direction as optic flow. For many amphibia, insects, and birds, the entire body partakes in the response (Rock, Tauber, & Heller, 1965; Reichardt & Poggio, 1976; Butterworth & Henty, 1991). As a rule, primates tend to exhibit more localized effects than other species so that eye, head, and whole-body movements tend to be controlled separately. Nevertheless, even human beings adjust their body posture when exposed to large moving visual surrounds (Dichgans, Held, Young, & Brandt, 1972;

Lee & Aronson, 1974; Delorme, Frigon, & Lagace, 1989). It is significant that the impact of optokinetic stimulation is largely contingent on a passive attitude of the observer, because the effect may be readily suspended or suppressed by active-looking or goal-oriented behavior. In humans and other species, OKR may be elicited by placing the observer within a rotating cylinder furnished with vertical stripes (see Fig. 5.03) or a random dot pattern. Stare–nystagmus can be initiated with instructions either to look straight ahead and not pursue the stripes, or to look straight ahead at an imaginary fixation point (Van Die & Collewijn, 1982; Van den Berg & Collewijn, 1988). In these studies, at a stripe velocity of about 10 deg sec^{-1}, eye velocity was only about 8 deg sec^{-1} and fell off from there at higher stimulus velocities. Thus, the gain (eye velocity/target velocity) of the slow phase of stare–OKR is considerably less than perfect. However, instructions to actively pursue or look at the stripes produced a gain of 1.0 at the slowest velocities and relatively higher gain at all other velocities as well (Van den Berg & Collewijn, 1986). Thus, in humans, active pursuit is necessary for accurate velocity matching.

In addition to its effects on the motor system, optokinetic stimulation has a unique effect on perception. Although one cannot be certain of the phenomenology of other species, in humans, an optokinetic pattern, such as vertical stripes on a surrounding cylinder or like that provided by large-screen movies, typically gives rise to the perception of self-motion, a sense of body movement termed *vection* (Dichgans & Brandt, 1978). The effect is especially compelling in flight simulators and in some virtual reality devices as well. It is a curious possibility that with each turn of the head or body and with each translational movement made with the eyes open, the accompanying optokinetic stimulation and the associated vection may actually contribute to the veridical perception of body movement. See Chapter 5 for a discussion of the possible role of eye movements in vection.

Smooth Pursuit

This type of eye movement is distinguished by the smooth record it produces in contrast with the jerky record of saccadic pursuit. It takes about 130 ms for the eye to begin pursuit after the target is set in motion (Robinson, 1965). Most accurate pursuit occurs for targets oscillating sinusoidally at about 1 Hz (Wyatt & Pola, 1983) and for constant velocity targets up to a maximum of about 100 deg sec^{-1} (Meyer et al., 1985). When these limits are exceeded, eye velocity increasingly lags image velocity and saccades tend to be used intermittently to keep up with the target. Although unlike saccades it cannot be initiated voluntarily without some sensory input or specialty training, SP seems to be influenced positively by factors such as predictability of target motion (Bahill & McDonald, 1983) and an active attentive attitude (Wyatt & Pola, 1987). Therefore, cognitive factors are thought to predominate, especially because accuracy generally suffers under passive viewing (Barnes & Hill, 1984). It is worth noting in this connection that cognitive impairment brought about by schizophrenia (Abel, Levin, & Holzman, 1992) or through drugs, such as alcohol and marijuana (Baloh, Sharma, Moskowitz, & Griffith, 1979), also degrades SP performance. For this reason, traffic police administer a simple test of the pursuit system, looking for multiple saccades as an indicator of inebriation. Also consistent with the influence of cognitive factors is the control exerted by motion perception and proprioception. For example, imagining the unseen hub of a wheel moving horizontally left and right while only two spots of light were visible opposite each other at the wheel periphery, as shown in Fig. 3.14(a), produced horizontal SP movements (Steinbach, 1971). Similarly, images that do not actually move across the retina but, as in the movies, give the appearance of real movement, readily evoke SP movements (van der Steen, Tamminga, & Collewijn, 1983). Furthermore, the motion of an unseen hand can be tracked in complete darkness (Jordan, 1970), and even the illusory movement of the forearm stimulated by muscle vibration of biceps or triceps, shown in Fig. 3.14(b), elicits an SP movement (Lackner, 1975). Thus, proprioceptive information from joint and muscle receptors is capable of driving the oculomotor system in the SP mode.

To the extent to which SP fails to stabilize the target image, the retinal slip spreads the image and degrades target recognition and identification. Successful stabilization, however, contributes not only to the cognitive processing of the target characteristics, but also to such perceptual characteristics of the target as its velocity or state of rest, as well as its radial orientation with respect to the observer. The various ways by which SP and other oculomotor systems contribute to perception are taken up directly in Chapter 4.

Vestibulo-Ocular Reflex and Its Adaptation

Perhaps the most significant function of the OKR stems from its association with the VOR, the third oculomotor system dedicated to the function of image stability. Unlike the two systems discussed previously, this system reduces image movement on the retina by stabilizing the motion of the eye in space with respect to an object also fixed in space. To accomplish this, a head rotation in one direction is accompanied by a compensatory eye rotation in the opposite

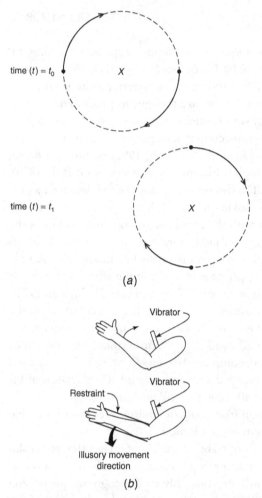

time (t) = t_0 X

time (t) = t_1 X

(a)

Vibrator

Vibrator

Restraint

Illusory movement
direction

(b)

Figure 3.14. *Perceptual and Proprioceptive Drives for SP.* (a) An unseen hub of a wheel marked by the *X* is seen as translating in continuous motion and therefore elicits smooth pursuit in the horizontal plane. Actual lights on the wheel periphery describe a totally different retinal path over time, and therefore they do not comprise the stimulus for horizontal SP (Steinbach, 1976). (b) Top: Vibration of the biceps muscle at about 100 Hz normally produces a tonic vibration reflex in the direction of the arrow due to the activation of muscle spindles (Hagbarth & Eklund, 1966) Bottom: When the forearm is constrained from moving, the biceps reflex continues unabated because the muscle cannot shorten to relieve tension on the spindles. The brain interprets this as attributable to muscle stretch as though the forearm rotated in extension through a nonphysiologic angle about the elbow joint (Goodwin, McCloskey, & Mathews, 1972a,b). When instructed to fixate the unseen index finger in darkness, eye movement recording indicated smooth following movements tracking the illusory trajectory of the index finger along a downward path (Lackner, 1975). Thus, at least some forms of proprioceptive information can be used to guide the output of the SP system.

direction, leaving the eyes directed toward the same spatial locus as before the head movement. In typical everyday life situations, as when nodding the head to signify "yes" or shaking to signify "no" or, in general, whenever the head rotates around any of the three spatial axes, compensatory eye movements occur. The reason for this response lies in the action of specialized underlying mechanisms found in the semicircular canals, that convey neural signals proportional to head velocity. These remarkable sensory structures are represented in Fig. 3.15(a–c).

In the ideal case, a head rotation in one direction would be accompanied by an opposite eye rotation of the same speed, thus yielding a gain (eye velocity/head velocity) of 1.0. In fact, when tested in the dark, one frequently finds a gain considerably less, perhaps around 0.6 (Barr, Schultheis, & Robinson, 1976). In the light, however, supplemented by the action of the OKR, moving images of the surroundings drive the eyes to follow them. Because optic flow is always opposite head rotation, both OKR and VOR operate to the same end. Under these conditions, a gain near 1.0 is common and gaze stability is achieved.

The OKR also provides a complementary function to vestibular stimulation when it comes to signalling continuous body motion. With continued rotational movement, the cupula, described in Fig. 3.15(c), settles down in about 20 to 30 sec. As a consequence, rotational velocities would cease to be signaled by the vestibular end organ were it not for the fact that the same cells in the vestibular nucleus that receive vestibular afference also are modulated by visual optokinetic stimulation (Waespe & Henn, 1977). Thus, as long as eyes are open, the sense of movement is sustained by the optokinetic system. See Chapter 6 for additional discussion of the concept of complementarity of function.

Finally, we note that the most significant of the various synergistic relationships between OKR and VOR actually derives from the adaptive nature of the VOR. Although it seems appropriate that, for example, a head movement around a vertical axis, which stimulates the horizontal canals, should also trigger a horizontal compensatory eye movement, the underlying neural circuitry is not unalterably dedicated to this end. Instead, one finds a great potential for adaptive plasticity. This includes the possibility of increased eye velocity, and therefore of increasing VOR gain in response to magnifying spectacles (Gauthier & Robinson, 1975) or prescription lenses (Collewijn, Martins, & Steinman, 1983); decreasing gain by viewing through minifying spectacles (Miles & Eighmy, 1980); reversing the direction of the VOR by wearing reversing Dove prisms (Gonshor & Melvill Jones, 1976); and altering the angular direction of the VOR (Schultheis & Robinson, 1981; Callan & Ebenholtz,

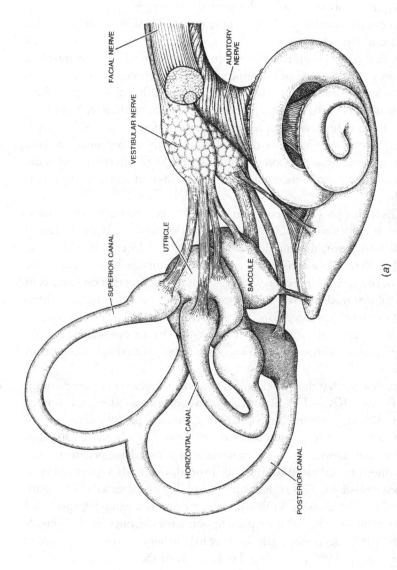

Figure 3.15. (a) *Labyrinth of Right Inner Ear.* (From Parker, 1980, The vestibular apparatus. *Scientific American, 243,* 118–134. Reprinted with permission of Patricia J. Wynne.) View is from the right, showing the horizontal and two vertical canals, and spiral cochlea, providing transduction of auditory signals. The entire structure lies within a space of 1 cm.³

(a)

FACIAL NERVE

AUDITORY NERVE

VESTIBULAR NERVE

SUPERIOR CANAL

UTRICLE

SACCULE

HORIZONTAL CANAL

POSTERIOR CANAL

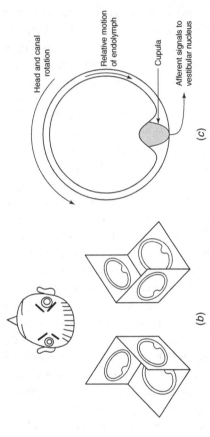

Head and canal rotation

Relative motion of endolymph

Cupula

Afferent signals to vestibular nucleus

(c)

(b)

Figure 3.15. (*Continued*) (b) *Stylized Representation of the Six Canals.* Above, seen from the top of the head, canals are shown behind the outer ear or pinna. Below, canals are represented in three orthogonal planes, mirror images of each other on either side of the median plane. Canals in the same plane act as a "push–pull" pair because rotation in any plane causes opposite changes in the resting-level neural output of each member of the canal pair. In any real head the canals are not exactly in orthogonal planes nor are they exactly parallel; therefore any head rotation produces some neural output from all three canal pairs. (c) *Canal Transduction.* A head acceleration to the left rotates the horizontal canal at the same velocity as the head, but the watery endolymph contents move at a slightly lower velocity, creating an inertial lag and pressure on the elastic walls of the cupula. The hair cells within the cupula respond in proportion to the shear force component acting on the cupula by increasing afferent neural output of the left organ while simultaneously decreasing output from the right paired canal. This difference signal is represented in various oculomotor nuclei and is used to drive the eyes in a plane approximating the plane of the active canals. In the present case, after a latency of about 14 ms (Lisberger, 1984) there occurs a leftward beating nystagmus, slow phase to the right, as well as a conscious sense of leftward head or whole body rotation.

63

1982; Khater, Baker, & Peterson, 1990) through the use of tilting prisms or other means. What this research suggests is that the direction and velocity of the visual optokinetic pattern may serve at least as a sufficient condition for adaptation of the VOR, provided that OKR and VOR occur concurrently (Miles & Eighmy, 1980); in other words, an optokinetic stimulus viewed with a restrained head, and therefore without canal stimulation, provokes no adaptive response.

There is a logical aspect to the critical role of OKR in adaptation that issues from certain characteristics of control systems based on feed-forward regulation. The canal–oculomotor nerve pathway is an example of such a feed-forward system that is consistent with the rapid 14 ms reaction time (Lisberger, 1984). These systems are rapid because no delay is required for feedback, but without feedback the canal does not "know" whether the eyes actually moved or in what direction and with what velocity. Such systems therefore are fast but ignorant, because they have no knowledge of the effect their behavior has on the environment they are intended to influence. To overcome this difficulty, feed-forward control systems usually are provided with look-up tables whose purpose is to create a control signal that is reasonably relevant to the real world. A heat-control system operating on these principles might have a clock that signals a quarterly change in seasons with a preprogrammed change in the amount of daily time the furnace stayed on. The system might then be correct for some typical season with some average temperature but would be inappropriate, except for chance, on a day-to-day or hour-to-hour basis. As look-up tables become more accurate by sending signals that represent the momentary state of the world, they become equivalent to feedback systems with attendant built-in delays. The role of the optokinetic system is analogous to that of a look-up table in that its function is to update the VOR so as to optimize gaze stability. In fact, because research has not yet distinguished between full-field optokinetic and foveal smooth-pursuit stimuli in regard to their respective capabilities to educate and retrain the VOR, it can be assumed for the present that both are capable of causing adaptation. Figure 3.16 represents an adaptive control system showing the updating function of the SP and optokinetic systems in relation to the VOR, which by itself has no feedback control.

Having observed that optokinetic stimulation and perhaps pursuit stimulation as well may provide the sufficient conditions for VOR adaptation (Ebenholtz, 1986), it may be wise to consider why these are not also necessary conditions. A very clear answer was provided by Melvill Jones, Berthoz, and Segal (1984) when they showed that an 11 percent reduction in VOR gain could be produced, in the absence of any visual stimulation, by training observers

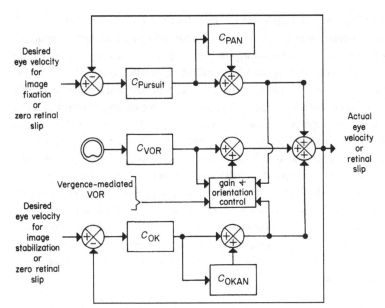

Figure 3.16. *A Self-Adaptive Control System for the Vestibulo-Ocular Reflex*
(*VOR*). (From Ebenholtz, 1986, Properties of adaptive oculomotor control systems
and perception. *Acta Psychologica, 63*, 233–246. Reprinted with permission from
Elsevier Science.) In traditional fashion for negative feedback systems, the output
(actual eye velocity) is subtracted from the input (desired eye velocity) to produce
an error signal, the negative of which forms the error-correcting quantity to be sent
through the system. Controllers, labeled *C*, operate on this signal by, for instance,
multiplying it or integrating it over time. Pursuit after-nystagmus (PAN; Muratore &
Zee, 1979) and optokinetic after-nystagmus (OKAN; Raphan, Matsuo, & Cohen,
1979) are velocity storage mechanisms that cause the continued SP and OKR after
the withdrawal of appropriate stimulation. They may be regarded as providing feed
forward signals that help to reduce the demand on the main error-correcting path.
Both OKR and SP ultimately feed into the gain and orientation controller that
serves to modify the VOR over time. Gain and orientation controller represents the
presumed source of adaptive plasticity. Note that as the VOR gain-and-orientation
controller gradually modifies the VOR output in conformity with either OKR or SP
error signals, the latter actually have less work to perform. This represents the
symbiotic relationship between feedback and feed-forward paths of adaptive
control systems; once the updating function of the negative feedback paths has
been fulfilled, they become quiescent because there is less demand to generate
error-correcting signals.

over a 3-hr period to maintain the gaze straight ahead of them while oscillating the head around a vertical axis. In effect, subjects practiced suppressing
the VOR by way of self-instruction and voluntary control of eye position.
Although the reduction in gain was only about one-third that obtained from

reversing prisms with visual stimulation present, for the same 3-hr period, nevertheless, visual stimulation cannot be claimed to be necessary for adaptation to occur. However, visual inputs probably are necessary for sustained and substantial levels of adaptation; Ebenholtz and Citek (1992, 1995) found that when voluntary (accommodative) vergence was sustained in darkness (see Fig. 3.10), subjects showed either no accommodative or vergence adaptation or low levels that rapidly decayed. In contrast, vergence and accommodation maintained via visual fixation on a near target produced relatively large sustained effects. Thus, as far as VOR, vergence, and accommodation are concerned, voluntary control produces decidedly lower and more transient levels of adaptation than that produced under reflexive control in the presence of visual stimulation. These outcomes suggest that the neural circuits employed in voluntary and reflexive oculomotor protocols are only partially overlapping and that volitional activities without concurrent feedback are inefficient at producing adaptation in oculomotor systems. Might the same principle apply to motor learning and adaptation in other domains?

Finally, one may note that the utility of the adaptive response extends beyond the laboratory context well into the "real" world where disease, aging, imperfect development, and medical interventions create inadequacies and asymmetries in the neural networks of the brain. Adaptive neurologic systems serve to redress these potential imbalances, initially through error-correcting negative feedback and finally by way of alteration in controlling parameters.

Yet another adaptive aspect of the VOR is represented in Fig. 3.17, which illustrates the functional utility that lies behind the modulation of VOR by the vergence system.

The Otolith–Ocular Reflexes

Signalling Mechanisms

Stimulation of the cupula of the semicircular canals [Fig. 3.15(c)] by radial acceleration is the first step in a signal shaping process, including encoding of head velocity that eventually triggers the compensatory VOR. Another set of biological transducers, the saccule and utricle, represented in Figs. 3.18 and 3.19, are located in small pouches at the junction of the canals [Fig. 3.15(a)]. These structures signal the *linear* acceleration of the head. One may experience linear acceleration in a vehicle that constantly increases its velocity along a straight path, but even more commonly, also when pitching or tilting the head while in the earth's gravitational environment, unless one is in free fall, as you might find

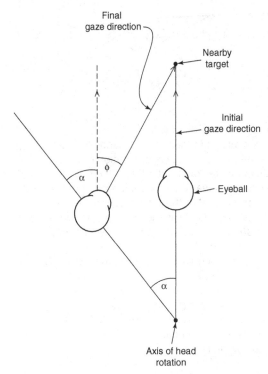

Final
gaze direction

Nearby
target

Initial
gaze direction

Eyeball

α

φ

α

Axis of head
rotation

Figure 3.17. *Near Targets Require Increased VOR Gain.* Because the axis of head rotation lies at the neck behind the eyes, they undergo a displacement as well as a rotation especially so in relation to nearby objects; for a target at optical infinity this differential approaches zero. Thus, in order to maintain gaze on a nearby target, the VOR must increase its eye velocity relative to that required by a distant target. This figure shows that when the head moves through an angle α, compensatory ocular rotation through the same angle would be insufficient to maintain target fixation unless an additional rotation angle ϕ were added. This is accomplished by increasing eye velocity relative to that of the head in response to an increased vergence signal (Post & Leibowitz, 1982), thereby effectively taking target distance into account. This is represented in Fig. 3.16 by the signal representing vergence-mediated VOR. Evidence also suggests a similar role for the accommodation system (Fisher & Sanchez, 1997).

yourself in a theme park or as an astronaut in earth orbit. It may seem curious that the two conditions, linear translation and head tilting, produce equivalent stimulation at the transducer level but, as explained in connection with Fig. 3.19, this is as it should be according to Einstein, who posited the fundamental physical equivalence between gravitational force and linear acceleration. There is nevertheless some "mischief" that results from this ambiguous encoding of

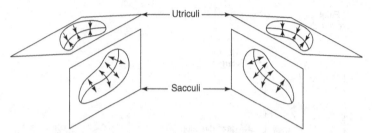

Figure 3.18. *Otolith Organs.* The saccule and utricle are mechano-electric transducers that signal linear acceleration and change in orientation relative to gravity. Although each organ occupies an area of about 2.5 to 3.5 mm², the saccule contains about 18,000 hair cells whereas the utricle supports about 34,000 hair cells (Rosenhall, 1972). They lie at the base of the semicircular canals and when viewed from the front occupy planes roughly as shown here. The curved line dividing each organ is known as the *striola*. It separates regions having opposite morphologic polarization axes, represented by the arrows. The polarization arises from the fact that the hair cells, protruding perpendicularly from their respective planes, respond maximally when hairs are displaced toward the kinocilium, shown in Fig. 3.19. In the utricle maximum response (depolarization) occurs in the direction toward the striola, but opposite the striola in the saccule. Only displacement of the stereocilia produces a nervous response (Hudspeth & Jacobs, 1979); thus the role of the kinocilium is not fully understood, nor for that matter is the function of the efferent supply shown in Fig. 3.19.

linear body movement and change in orientation. One type of problem arising from the identity of stimulation during forward translation and backward pitch, associated with aircraft carrier take-offs, has been described in conjunction with Fig. 3.19. A similar problem described later involves the association of otolith stimulation with body translation during space orbit, but first it is necessary to describe the basic otolith–ocular reflexes on earth.

Ocular Counterrolling and the Doll Reflex

In a gravitational environment, a tilt of the head toward one shoulder or a shift of the entire body as when lying on one's side provokes a counterrolling (torsion) of the eyes around the line of sight (Miller, 1962). The rolled position of the eyes is sustained for as long as the head is tilted; therefore, the canals are not implicated. The eye movement, thought to be driven primarily by the utricles (Diamond & Markham, 1983), is compensatory in the sense that a clockwise head rotation produces a counterclockwise ocular rotation, but the gain (eye rotation/head rotation) is quite low: at a 90 deg tilt, right ear

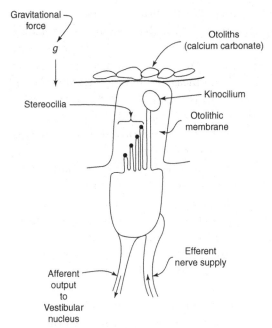

Gravitational force

g

Stereocilia

Otoliths (calcium carbonate)

Kinocilium

Otolithic membrane

Efferent nerve supply

Afferent output to Vestibular nucleus

Figure 3.19. *The Equivalence of Linear Translation and Head Tilt.* A stylized cross section of a hair cell mechanism with only four stereocilia are shown whereas actually more than 50 hairs per cell are common. Imagine the hair cells distributed over the sacculi and utriculi with the hairs perpendicular to the pictorial planes, whereas the otoconial layer, containing the calcium carbonate crystals, is parallel to this plane. If the hair cell is rotated clockwise so that gravitational force acts along the otoliths, this force is transmitted through the otolithic membrane to the kinocilium and the distal tips of the hair bundle to produce an increased afferent output. A force acting away from the kinocilium [to the left in Fig. 3.19] causes hyperpolarization, a reduced response, whereas intermediate angles yield output proportional to the cosine of the force angle (Hudspeth, 1983). These organs probably encode head position relative to gravity and signal antigravity muscle reflexes and oculomotor responses such as torsion and the doll reflex (see Fig. 3.20 and Chapter 4).

It is significant that a constant linear acceleration to the left because of the greater inertial lag of the otoliths in relation to the hair bundle displaces the stereocilia toward the right, thereby increasing afferent output. Thus, in the example given, a clockwise tilt in a gravitational field and a linear acceleration to the left may, in principle, produce identical afferent neural output. This equivalence may have been at the root of airplane disasters when aircraft carrier pilots, on accelerating at take-off, felt they were excessively pitched upward. Correction from this illusory attitude promptly caused an irrecoverable dive into the ocean (Cohen, Crosbie, & Blackburn, 1973).

down, only about 9 deg of compensatory eye rotation may be expected (Bucher, Mast, & Bischof, 1992). This low gain indicates that the torsional reflex is barely functional. Consistent with expectations from the equivalence of otolithic stimulation due to left or right head tilt and lateral translation, human torsional eye movements have been observed and measured during linear oscillation along the interaural (left–right) axis (Lichtenberg, Young, & Arrott, 1982), where they certainly cannot contribute to gaze or image stability.

As one might expect from an oculomotor system driven by mechanisms sensitive to gravito-inertial forces, the torsional positions of the two eyes depart from their normal 1-*g* positions during the 0-*g* phase of *parabolic flight* (Money et al., 1987). In what has become a conventional procedure, NASA researchers typically use a KC-135 aircraft as a flying laboratory. To achieve a 0-*g* state, the pilot approximates a parabolic trajectory, at the peak of which, the aircraft pitches downward and enters a brief period (about 25 s) of free fall. The aircraft then gradually pitches upward and accelerates, thereby increasing gravito-inertial force, to achieve about 25 s of 1.8-*g*, then decelerating until the peak of the parabola is reached. The process usually is repeated about 20 times on the outward bound leg and again 20 times on the return portion of the trip. Nausea and emesis are frequent-flyer rewards on these flights.

Because the 0-*g* state characterizes the space environment of astronauts (see later, the discussion of adaptation in space), it is most likely that similar ocular torsional displacements also occur in earth orbit, where it is thought that they contribute to space motion sickness (Diamond & Markham, 1991). Accordingly, it may not be mere coincidence that, when Balliet and Nakayama (1978) succeeded in training their earthbound subjects to produce voluntary torsion, which, surprisingly, yielded rotational amplitudes as high as 30 deg, subjects experienced nausea and other motion sickness symptoms. (See Chapter 5 for a further discussion of the role of ocular muscle traction in motion sickness.)

Finally, it is worth noting that torsion may thus join accommodation as members of a class of ocular reflexes that, with proper feedback, may be placed under voluntary control.

In a fashion somewhat analogous to ocular counterrolling, when the head is pitched back, as when looking up, or when the whole body is pitched back in a supine posture, the downward acting ocular muscles are stimulated reflexively, similar to the movement of counterweighted doll eyes. This produces a bias in the position of the eyes that the subject believes to represent a straight ahead gaze (i.e., the subjective straight-ahead eye position shifts downward with backward pitch; Ebenholtz & Shebilske, 1975; Citek & Ebenholtz, 1996). Peak shift with backward tilt occurs around 90 deg of pitch, as shown in Fig. 3.20, and

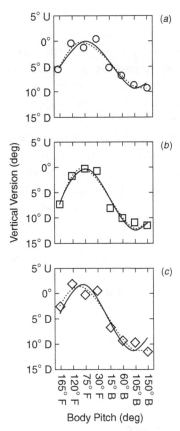

Figure 3.20. *The Doll Reflex with Body Pitch.* (From Citek & Ebenholtz, 1996, Visual and horizontal eye displacement during static pitch and roll postures, *Vestibular Research, 6,* 213–228. Reprinted with permission of I. O. S. Press.) Subjects were placed at pitch angles throughout 360 deg of space in 45-deg increments. When they signaled that they were looking straight ahead, an infrared video image was made of eye position. Data represent deviation from a baseline at zero deg body pitch. In (a) pure pitch is represented, whereas (b) and (c) show the reflex with the body pitched and also rotated 45-deg clockwise and counterclockwise, respectively. The dotted line represents the best fit third-order polynomial function, whereas the solid line derives from the best fit rational function based on trigonometric expressions for the shear components of force operating on the hair cells of the utricles and saccules during pitch and roll tilt. The general form of this function is:

$$Z = D \sin \rho + C \, \sin(\phi + 30)\cos \rho + B \, \cos \phi \cos \rho + A$$

where z is the vertical version, ρ and ϕ represent roll and pitch angles, respectively, and A, B, C, and D represent coefficients consistent with the best fit function. The coefficients of determination R^2 showed that from 83 to 93 percent of the variance in vertical eye position was accounted for by the rational function. On the axes, B and F represent backward and forward body pitch, respectively, and U and D represent up and down, respectively.

is under about 10 deg; therefore of the same order of magnitude as the counterrolling reflex. Likewise, linear acceleration in the naso-occipital direction (straight ahead or backward with head upright) induces vertical responses that are nonfunctional at low frequencies of oscillation (0.5 Hz, Paige & Tomko, 1991a,b). In the latter study, their subjects (squirrel monkeys) showed signs of a remarkable oculomotor mechanism, a gaze-dependent compensatory response. They found, for example, that during forward movement in darkness when initial gaze direction was straight ahead, no response was made, whereas when initially gazing to the left, a further leftward response occurred, and likewise the eye rotated rightward during right gaze and upward or downward during up and down gazes, respectively. The convergence angle and eccentricity of initial gaze also heavily influenced the subsequent response. No comparable data are currently available on humans, nor is it known what influence the immobilization of the head had on these responses. Nevertheless, given the strong conditional nature of these oculomotor responses, it seems likely that they were shaped by the optic flow patterns to produce functionally valid linear vestibulo-ocular reflexes (LVOR) conditional on initial gaze direction. Such conditional changes have indeed been demonstrated in the case of the gain of the horizontal VOR associated with up and down gaze (Shelhamer, Robinson, & Tan, 1992).

Otolith–Ocular Adaptation in Outer Space

In order to consider how the otolith–ocular reflexes behave in outer space, it is important to note that the distance from Earth to an orbiting space ship, about 300 miles, is much too short to reduce the force of gravity by more than about 15 percent. Nevertheless, the gravito-inertial force on astronauts is effectively zero because they are in a condition of perpetual free fall, owing to the balance achieved by the centripetal g-force and the tangential acceleration that maintains ship and crew in circular orbit. Because all objects, regardless of mass, fall at the same rate in a gravitational field, the hair cells of the otolith organs are not affected by the overlying otoliths, regardless of head angle assumed. Thus, with head tilt, as far as signals from the utricle and saccule are concerned, except for the spontaneous, resting-level neural output and response to linear acceleration, the brain in space orbit is essentially de-afferented, as though the otolith-organ portion of the eighth nerve had been cut, while sparing the signals from semicircular canals. A similar unloading of the utriculi (but not the sacculi) occurs when upside down on Earth, and may be related to the frequent reports of subjective inversion by astronauts in orbit (Kerwin, 1977) and in the zero-g phase of parabolic flight (Graybiel & Kellogg, 1967). Under these conditions, apparent body orientation is heavily influenced by vision and proprioception (Lackner & Graybiel, 1983).

As noted previously, although otolith organs do not respond to head tilt in space orbit, they do respond to linear acceleration. This arises out of the special type of human locomotion required in space. Here, walking is not possible. Instead, one is launched into space each time the toe pushes off. Hence, astronauts tend to glide and dive (glive?) toward their destination. These sequences of transitional acceleration and deceleration trigger transient otolith activity much more frequently than is the case on Earth. The combination of an optic array signalling translation together with otolith activity may be a sufficient adaptive stimulus for a change both in the central interpretation of otolith stimulation and in the type of eye movement normally triggered. This has been termed the *tilt-translation reinterpretation hypothesis* (Young et al., 1984; Parker et al., 1985). Accordingly, one finds on return to Earth reports by astronauts that head tilt elicits a sense of vection with translational components (Parker et al., 1985), and many reports of a reduction in ocular counterrolling after space flight (e.g., Vogel & Kass, 1986; Dai et al., 1994). Thus, it appears as though the adaptive neural networks of the vestibular cerebellum are extremely active in outer space.

An Overview of Oculomotor Systems

Categories of Oculomotor Function

A wide variety of interesting ways are available by which to categorize and study eye movements, including ontogenetic (Aslin & Dumais, 1980) and phylogenetic (Walls, 1962; Hughes, 1977) approaches. A different but equally pragmatic set of categories is represented in Table 3.00 according to nominal stimulus conditions and apparent visual function. The classification may have some heuristic pedagogic value, but it is not perfect because many system attributes actually are shared. For example, typical saccades occur during the quick phases of VOR and OKR (Ron et al., 1972); saccades play a significant role in asymmetric vergence (Enright, 1992); and slow drifting eye movement, not just saccades, can be used to maintain foveation (Steinman, Cunitz, Timberlake, & Herman, 1967). These and many other observations provide ample evidence for the extensive sharing of structure and function across the various oculomotor systems. The present piecemeal approach reflects a convenience of analytic research, soon to give way to more integrative studies

Table 3.00. *Oculomotor systems grouped by stimulus condition and primary function*

Types of stimulation	Oculomotor system	Primary function
Optical, perceptual, and self-instruction	1. Vergence 　　Disparity vergence 　　Accommodative vergence 　　Pattern-driven vergence	Single vision
	2. Accommodation 　　Blur-driven accommodation 　　Convergent accommodation 　　Patterndriven accommodation 　　Voluntary accommodation 3. Fixation reflex (saccades)	Clear and detail vision
	4. Foveal pursuit 5. Optokinetic response (OKR)	Gaze and image stability
Inertial	6. Vestibulo-ocular reflex (VOR) 7. Otolith-ocular reflex 　　Doll reflex 　　Ocular counterrolling	

reflecting system interactions. For example, Schor, Lott, Pope, and Graham (1999) investigated saccades that occur immediately after a blur stimulus and found them to reduce the latency of focusing and increase the speed of both accommodation and accommodative vergence. Likewise, saccades have been shown to facilitate changes in disparity vergence (Enright, 1984). Thus, the next phase of research has already begun.

4

Oculomotor Factors in Perception

Introduction

In 1865, Claude Bernard, the great French physician scientist regarded as the father of experimental medicine, insisted on what he termed the *method of counterproof.* This is embodied in the Latin phrase *sublata causa, tollitur effectus*, meaning "remove the cause, eliminate the effect." Bernard encouraged his fellow physicians to seek for truth by evaluating contradictory facts that challenged their favorite hypotheses, and emphasized that "the only proof that one phenomenon plays the part of cause in relation to another is by removing the first, to stop the second" (Bernard, 1865/1957, p. 56).

Some of the evidence for the role of oculomotor systems is of the type advocated by Bernard, where, for example, as discussed later, vestibular nerve damage was shown to result both in the elimination of the vestibulo-ocular reflex (VOR) and the simultaneous obliteration of perceptual stability. Mostly, however, research reflects a strategy whereby instead of eliminating a given oculomotor system, a system parameter has been altered and changes in perception are subsequently measured. So long as the method used alters only the oculomotor system and is not also capable of directly changing perception, the method is a powerful one for evaluating causal relationships between perception and oculomotor systems.

Topics covered include:

Direct and mediated spatial properties of oculomotor systems.
Where the self lives and works (the ego-center and the cyclopean eye).

Illusions that result from balancing voluntary against reflexive eye movement control signals.

Ghosts, afterimages, and lightning.

Information Extraction and Perceptual Attribution

Given the large number and complexity of systems devoted to keeping images clear and fixed on the retina, the easy implication to draw is that information extraction is thereby optimized; certainly, recognition and identification would be degraded by the smear of moving or the blur of out of focus images. However, the same oculomotor mechanisms also contribute to an inverse process of attribution whereby spatial properties of perception, such as apparent direction and distance, are assigned to the objects signaled by their retinal images. Indeed, if, for example, egocentric spatial location were not one of these attributive properties, then having identified an object as dangerous while neither registering nor perceiving its location with respect to the observer would surely provoke dysfunctional behavior. Of course, all perception is attributive in the sense that all perceptual properties arise out of some form or pattern of brain stimulation, and as such do not have an independent existence in the external world. It may be surprising, however, to learn how extensively oculomotor systems contribute to this process.

Perceptual Properties of Oculomotor Systems

What are the spatial properties that are rendered salient by virtue of their underlying oculomotor systems? There appears to be two categories of properties, direct and mediated, distinguished by the degree to which the property in question is more or less directly related to some aspect of oculomotor behavior. These distinctions are brought out demonstratively later. Among the direct set are such attributes as apparent radial direction, apparent horizon or eye level, apparent target velocity (including movement path), and apparent distance. Indirect attributes include frontal plane orientation, apparent size and depth, and apparent vertical in the median plane.

Radial Egocentric Direction

Imagine one or both eyes directed toward a point of light in an otherwise darkened environment. Where is the light seen in relation to the observer?

To answer this, we must first note that the human body is segmented so that different body parts may be used as frame of reference. For example, in principle, we may judge whether a target is left or right of the fovea or gaze direction without even registering where in space one is looking. Such a pure foveocentric (or oculocentric) judgment may actually never occur because a headcentric judgment that seems much more functional usually can also be made. To make a headcentric judgment or, assuming the self resides in the head, an egocentric judgment, it is logically necessary to register the location of the image in relation to the fovea and the fovea (or eye) in relation to the head. Then, because objects tend to be seen along direction lines that run from the retinal image through the ocular nodal point to the object, knowing the eye-in-head position determines the location of the direction line relative to the head and therefore the headcentric position of an object. For example, an object whose image is 10 deg to the right of the fovea in an eye 10 deg to the right of straight ahead is appreciated as 20 deg to the right of the observer, or more accurately, to the right of the median plane of the observer's head. By a similar line of reasoning, a body frame of reference, such as the median plane of the trunk, may be used to specify apparent object position by registering the position of the image with respect to the retina, the eye with respect to the head, and the head in relation to the trunk.

Figure 4.00 represents an egocentric coordinate system, based on the version and vergence systems, that encodes angular orientation in the vertical and horizontal planes, as well as the distance to the target. Although there is little doubt that fixating a target imbues it with an egocentric orientation and distance, exactly how eye position is signaled is not fully understood. Two alternatives are generally accepted and both are based on extraretinal information. In one case, *inflow* to the brain from transducers in the extraocular muscles, such as muscle spindles, is presumed to encode information about muscle length and eye position (Sherrington, 1918; Lukas et al., 1994). In the second case, *outflow* or efference used to innervate extraocular muscle is presumed to be copied and made available to encode eye position (von Holst, 1954; Sperry, 1950). Whichever is valid or even if a hybrid theory is correct (Matin, 1976), the perception typically is bipolar in the sense that the observer is aware of both the direction of his/her gaze and also of the egocentric orientation of the target.

The Egocenter and the Cyclopean Eye

Sensing the direction of an object implies the existence of a center from which direction lines emanate. This center is the origin of the coordinate system

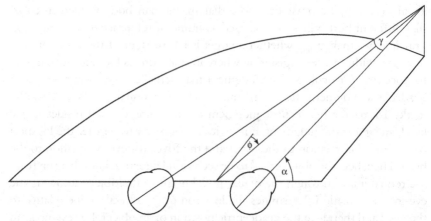

Figure 4.00. *A System of Egocentric Angular Coordinates.* (From Ebenholtz, 1984, Perceptual coding and adaptations of the oculomotor systems. In L. Spillman & B. R. Wooten (Eds.), Sensory experience, adaptation, and perception: *Festschrift for Ivo Kohler*, Chap. 19. Reprinted by permission of Lawrence Erlbaum Associates, Inc.) The convergence angle γ encodes distance, whereas the angles α and ϕ encode elevation and horizontal orientation, respectively. The latter derive from the version or conjugate system of eye movements. When the eye does not fixate the target of interest, its egocentric orientation may nevertheless be registered provided that (1) eye position in the head is known, and (2) the angular deviation of the target image relative to the fovea also is known (Morgan, 1978).

Sensing the location of an object in relation to some part of the self is a precursor for action. One may then direct the hands toward it, walk toward or runaway from its locus, direct one's voice properly, or otherwise send signals toward the object. Thus, egocentric localization is of central importance for animal (and robot) behavior, and therefore it would not be surprising to find redundant systems evolved for the task of spatial orientation. Although psychologists have devoted a great deal of intellectual effort to devise theories of how the dynamic image on the retina of a moving observer may encode orientation information (Gibson, 1966; Hildreth, 1992; Perrone, 1992), research supports the integration of this approach with a corresponding role for oculomotor systems (Royden, Crowell, & Banks, 1994; Bradley et al., 1996).

shown in Fig. 4.00 and the point between the eyes from which the world is perceived (Fig. 4.01). These ideas were developed by Hering (1879/1942) and expressed in the notion of a hypothetical cyclopean eye, as shown in Fig. 4.01(c and d).

The principle represented in Fig. 4.01(d) is that eye position must be registered in the brain in order to associate an egocentric direction with any particular point in a retinal image. One implication of this fact is that the pattern of brain stimulation produced by a retinal image is thoroughly ambiguous

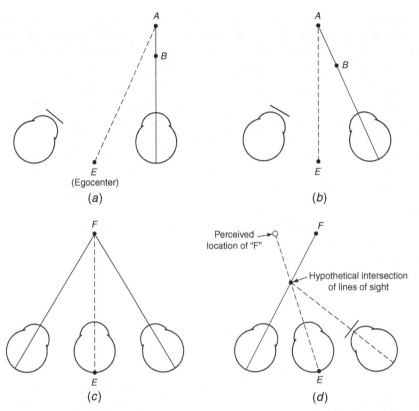

Figure 4.01. *The Egocenter and Cyclopean Eye.* It is curious and significant that given any point or any visible object *A* and a second point *B* visually aligned with it, then a line going through the two points will appear to be directed toward an egocenter, usually somewhere between the two eyes. If the second point in the example is taken to be a fingertip, then a line joining the target point and fingertip is perceived as pointing toward the self and, appropriately enough, the fingertip points toward the target. The phenomenologic fact of an egocenter in line with a pointing finger and target would seem to justify the common belief in an "I" that is doing the pointing and seeing, as if from a single point midway between the eyes. In Fig. 4.01(a and b), the target and finger actually are aligned with the fovea of the right eye, but in these and all other cases, target and finger are experienced as directed back toward a point near the median plane of the head (Jones, 1965). The concept of egocenter thus expresses the common experience of viewing the world from a single vantage point, no matter what the eye position within the head.

Figure 4.01(c) demonstrates Hering's concept of the cyclopean eye under binocular viewing. Even though each eye views the world from a different vantage point, as long as corresponding points are stimulated, a single object is seen, but in what direction? Hering's rule for binocular directions was that the fixation target would be located on a plane passing through the egocenter *E* on one pole, and a point in space *F* as the other pole, corresponding to (*Continued on next page*)

with respect to directionality unless eye position information is added. It also follows that the identical locus of brain stimulation would signal distinctly different egocentric directions if the stimulation corresponded to different eye positions. An application of this concept is represented in Fig. 4.02.

It seems clear that the operational capability of the brain may be widely expanded by the use of conditional eye position inputs, much like the capability of the gold key on a hand calculator to increase the functions associated with any given button. This represents a complication for the development of visual prostheses (Brindley & Lewin, 1968; Sterling, Bering, Pollack, & Vaughan, 1971; Normann, 1995) by direct stimulation of the visual cortex because, in order to convey directionality, some way would have to be found to introduce head-referenced information about the surrogate-eye replacement for the real eye. Not surprisingly, a blind patient in a prosthesis study reported, "that the phosphenes [patches of light] elicited by direct cortical stimulation appeared to shift in conformity with movements of the eyes" (Sterling et al., 1971, p. 70).

Perceptual Effects of Biased Horizontal Eye Position Information

a) Apparent visual direction. Early in the study of adaptation to prism-induced optical distortions psychologists found that a short period of wearing wedge-shaped prisms that optically displaced the visual scene (see Chapter 2) also produced oculomotor shifts (Kalil & Freedman, 1966; McLaughlin & Webster, 1967). These in turn were associated with the shifted position of a visual target judged to be straight ahead when viewed without prisms. Therefore, exposure to laterally displacing prisms constitutes a method of biasing eye-in-head information and produces changes in apparent egocentric direction. Subsequent studies have shown that maintaining fixation on a point-target in asymmetric posture without prisms also suffices to bias the straight ahead eye posture as well as apparent visual direction (Paap & Ebenholtz, 1976).

Figure 4.01. (*Continued*) the intersection of the two lines of sight. Figure 4.01(d) represents the application of this rule when a covered eye rotates into an esophoric position. Deductions from Hering's rules predict an illusory shift of the fixation target *F* away from the covered eye, which has been found (Park & Shebilske, 1991; Ono & Gonda, 1978). One implication of these results is that the registered location of both eyes is taken into account when the brain computes apparent egocentric direction.

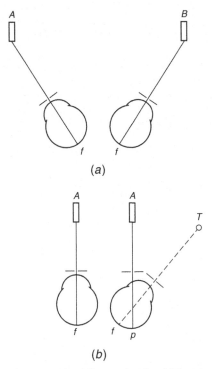

(a)

(b)

Figure 4.02. *Change in Visual Direction and Stroboscopic Movement.* As shown in Fig. 4.02(a), luminous bars *A* and *B* flashed on and off stimulating the fovea *f* alternately while the subjects attempted saccades to the left and right in synchrony with the flashes (Rock & Ebenholtz, 1962). Most subjects who could accomplish this saw a single bar moving alternately from left to right and back again in stroboscopic motion. Others reported seeing two bars alternately flashing on the left and right. In both cases, the bars were seen in totally different locations even though they stimulated very nearly the *identical* foveal–cortical locus.

In Fig. 4.02(b), subjects made saccades, monocularly, between the flashing bar at *A* and the steady fixation target at *T*. When fixating *T*, the bar at *A* flashed again, but this time stimulating the peripheral retina at *p*. Although the rate of saccadic alternation and flashing were the same as that in Fig. 4.02(a), none of the subjects who experienced motion in that context reported any movement at all under the conditions of Fig. 4.02(b). The authors concluded that although the flashing bar alternated between very different retinal–cortical loci at *f* and *p*, the absence of change in apparent direction of *A* caused the failure of apparent movement in the conditions of Fig. 4.02(b). Therefore the change in egocentric direction of the bars, as in Fig. 4.02(a), was necessary for apparent movement to be seen, even though in that case retinal–cortical loci were unchanged.

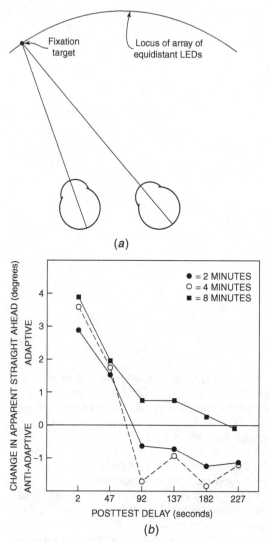

(a)

(b)

Figure 4.03. *Change in the Visual Location of a Target Judged to Be Straight Ahead as a Function of Test Interval after Maintained Fixation at 32-deg Eccentricity.* Subjects maintained fixation on a small LED placed 32 deg left or right of straight ahead, as represented in Fig. 4.03(a), for 2, 4, or 8 min. Before and 2 sec after the fixation period, subjects chose from a horizontal array of equidistant LEDs a target that appeared straight ahead. Testing continued in darkness at 45 sec intervals up to 227 sec, thereby permitting an assessment of the rate of decay of the aftereffect. These data are shown in Fig. 4.03(b). (From Paap & Ebenholtz, 1976, Perceptual consequences of potentiation in the extraocular muscles: An alternative explanation for adaptation to wedge prisms. *Journal of Experimental Psychology, Human Perception and Performance, 2,* 457–468. Copyright 1976 by the American Psychological Association. Adapted with (*Continued on next page*)

Therefore, even though it could be argued that a prism-displaced visual scene contained information for *both* the change in eye posture and the shift in apparent visual direction, there is little doubt that an oculomotor shift induced in the absence of visual information, itself, unless it is somehow unrecognized by the nervous system, entails a shift in perceived direction of a visual target.

Figure 4.03 shows the shifts in apparent visual direction in a study by Paap and Ebenholtz (1976) in which subjects first maintained asymmetric binocular fixation on a target placed 32 deg left or right of straight ahead. It was inferred that overall, an oculomotor shift ranging from about 3 to 4 deg was produced in the direction in which the gaze was previously maintained which, in turn, caused a visual shift of the same amount. Similar direction aftereffects have been recorded as pointing errors after sustained leftward head rotation, as well as after combinations of head and eye rotation (Ebenholtz, 1976). Thus, apparent visual direction may be influenced by both eye- and head-position bias.

Another direct test of the role of oculomotor systems in the perception of direction was carried out by Skavenski, Haddad, and Steinman (1972). Unlike adaptation protocols that bias eye position as a result of aftereffects of previously sustained ocular postures, these authors biased eye position by physically increasing the load on the eye, thereby increasing the muscle tension required to move and to maintain fixation. Figure 4.04 represents the method and task of the subject. Results clearly showed that in order to perceive the egocentric direction of an object whose image is represented on the fovea, the system requires the registration of eye position information that, when biased, causes related shifts in apparent target direction. The authors believed these shifts to reflect primarily changes in outflow to the extraocular muscles, but because the pulled eye was required to maintain fixation against the force of the pull, inflow probably also changed as well.

The idea of investigating egocentric orientation by physically manipulating the eye position in its orbit actually is traceable to Descartes (1664/1972).

Figure 4.03. (*Continued*) permission.) It is clear that direction aftereffects of eccentric ocular fixation occur, and that they become less transient and more truly plastic in the sense that the effect is more likely to remain with increasing exposure time. It is also possible that continued testing actually served to hasten the decay process.

The adaptation that results from sustained eccentric gaze may be interpreted as a shift in the resting level of the oculomotor system that controls the primary position or straight-ahead gaze direction.

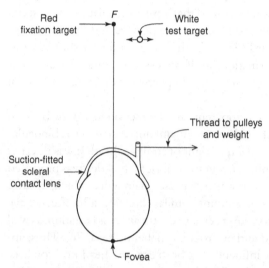

Figure 4.04. *Method for Manual Rotation of the Eyeball.* (Skavenski 1972; Skavenski et al., 1972). In a type of heroic experiment, a tightly fitting scleral contact lens mounted with a stalk was worn by Alexander Skavenski and Robert Steinman. By systematically loading a string tied to the stalk while maintaining fixation of the right eye on a red target light *F*, a compensatory increase in eye muscle tension was engendered. Because in the figure the weight on the string pulls the eye toward the right, compensatory innervation is as though the eye were pointing toward the left, even though its direction remains straight ahead, in line with a midsaggital plane passing through the body. Changes in the position of the white target when set to appear straight ahead were taken as the measure of the change in apparent position of the red target because the two were set in correspondence when no load was present. Results showed that the apparent position of the red fixation target shifted opposite the side of the load in the direction of the compensatory innervation.

It is noteworthy that when the subject was allowed to view the red fixation target with the *left* eye while the occluded right eye was manually rotated, the "red point appeared stationary no matter what was done to his right eye" (p. 289). Using a similar procedure, the same null result was reported by Gauthier, Nommay, and Vercher (1990). However, when these authors had their subjects point at a target with their unseen hand, large and significant shifts were obtained. The target, viewed with the nonloaded eye, seemed to be displaced in the same direction in which the covered eye was pulled, although not to the same extent. Thus, the position of both eyes, including the displaced, covered one, contribute to apparent target direction, but evidence for or against this conclusion seems to be contingent on the use of either manual or purely visual measures, respectively.

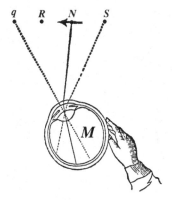

Figure 4.05. *Descartes' Eye-Press Paradigm.* Descartes hypothesized that errors in perception would occur when the muscle forces driving the eyes or fingers were interfered with by outside constraints. In that case, he suggested that the brain would register some intermediate position. For example, of the eye-press paradigm, he wrote "if eye *M* is turned away by force from object *N*, and arranged as if looking toward *q*, the soul will judge that the eye is turned toward *R*. In this situation rays from object *N* will enter the eye in the way that those from point *S* would do if the eye were in fact turned toward *R*; hence it [the soul] will believe that this object *N* is at point *S* and that it is a different object from the one being looked at by the other eye" (Descartes, 1664–1972, *Treatise of Man*, p. 65 in Hall translation). Descartes use of the religious term *soul* can be understood in contemporary times as referring to the conscious representation of certain brain states.

Figure 4.05 represents Descartes' eye-press paradigm by which he sought to demonstrate that the apparent location of an object in space is contingent on both the position of its retinal image and the brain's registration of the location of the eye in orbit. Descartes offered the hypothesis that the brain registers an eye position intermediate between the initial fixation position (*N* in Fig. 4.05) and the final position caused by the finger pressure (*q* in Fig. 4.05). In Descartes' figure, if the eye were registered as having turned toward *R*, short of *q* then the displaced image of *N* would be perceived as if it were at position *S*. Simply put, the eye-in-orbit position of the pressed eye is misregistered as being to the right of its true location. Accordingly, all images on the retina of such an eye lead to the perception of objects as being also to the right of their true locations.

Modern studies of eye-in-orbit information (e.g., Bridgeman & Stark, 1981; Stark & Bridgeman, 1983) strongly support the premise that efference sent to the extraocular muscles is monitored by certain parts of the brain and represented in conscious experience (Helmholtz, 1910/1962, Vol. 3; Sperry, 1950; von Holst, 1954). Thus, in so far as the eye-press paradigm produces a biased

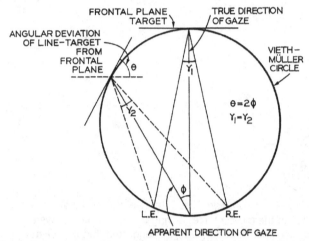

Figure 4.06. *Model of the Conditional Nature of the Apparent Frontal Plane (AFP).* (From Ebenholtz & Paap, 1976; Further evidence for an orientation constancy based upon registration of ocular position. *Psychological Research, 38*, 395–409. Reproduced by permission of Springer-Verlag GmbH & Co. KG.) A Vieth–Müller circle drawn through the nodal points is an equiconvergence surface (see section on horopter in Chapter 3). Because binocular disparity may be defined as the difference in convergence angles, the Vieth–Müller circle also represents a zero-disparity locus for any pair of points on the circle. In contrast, pairs of points, at least one member of which is *not* on the circle, produce a gradient of disparities as the target line or plane departs more and more from the circle. It is perhaps counterintuitive that a flat surface, such as a conventional wall viewed frontally, actually produces such a gradient of disparities because points on the same wall do not appear to be at differing depth intervals. Thus, our conception of flatness entails a disparity gradient.

In order to appreciate why the Vieth–Müller circle may serve as a model for the AFP, imagine the effect of sliding the frontal-plane target, shown in Fig. 4.06, along the Vieth–Müller circle through an arc of about θ deg. This leftward displacement, which is the equivalent of an angle at $0.5\,\theta$ or ϕ deg at the egocenter, is accompanied by a reorientation of the line by θ deg out of the frontal plane. A Gedanken experiment shows that by keeping the line tangent to the circle, the disparities remain unchanged. This may be demonstrated by producing an afterimage of the line while viewing straight ahead at the frontal-plane target. If one then shifts the direction of gaze to the left, it is apparent that the pattern of binocular disparities is not altered because these are fixed by the afterimage. Because one could not discriminate between this afterimage and a real line that is shifted and rotated, it follows that if the model is correct, with retinal input fixed, what we consider to be the AFP depends essentially on ocular gaze direction. The effect of a leftward gaze bias on AFP is demonstrated in Fig. 4.06. Here, fixation actually is straight ahead at the center of the frontal-plane target, but the observer's apparent direction of gaze is assumed to be to the left (*Continued on next page*)

efferent pattern, Descartes' formulation may be regarded as essentially correct and remarkably modern. Careful experiments by Bridgeman and colleagues (e.g., Ilg, Bridgeman, & Hoffmann, 1989) have shown the eye-press to produce an actual displacement or translation of the pressed eye that is compensated by an efferent command to rotate the eye in the opposite direction so as to ensure continued accurate fixation. Central registration of this efferent command (signaling that the eye has moved opposite to the finger pressure) is sufficient to account for an illusory shift in target position, opposite in direction to the eye-press force. An analysis by Rine and Skavenski (1997) complicates matters a bit by suggesting that the eye-press paradigm actually results in a force pattern that not only displaces the eye, but also introduces a rotary force. Both forces would have to be compensated for by an outflow signal in order to maintain target fixation by the pressed eye. Because the outflow signal is sent to both eyes (see Chapter 3), by monitoring the movement of the nonpressed, covered eye, Rine and Skavenski found evidence for the required outflow signal and also for a change in apparent target position in the same direction as the compensatory eye movement. The brain apparently does not discount the compensatory outflow signal as being due to the voluntarily initiated eye press, but treats it as though the target were shifted in space. (See Chapter 5 for further observations on the eye-press paradigm.)

b) Apparent frontal-plane orientation. The apparent frontal plane (AFP) is a plane that appears to be parallel with the forehead or shoulders, as when stationed directly in front of a wall. It may be surprising that the apparent slant of this plane should be governed by oculomotor factors, especially by the registration of lateral gaze direction, but evidence suggests that the visual input in the form of the image or the pattern of binocular disparity must be interpreted together with the gaze angle in order to perceive a surface with an unambiguous slant angle (Ebenholtz & Paap, 1973, 1976; van Ee & Erkelens, 1996). The somewhat complex nature of this phenomenon justifies considering the AFP as one of the instances of indirect or mediated effects of oculomotor factors on perception. A geometric model of the relationship between gaze direction and AFP is represented in Fig. 4.06. If perception follows the model, then when

---

Figure 4.06. (*Continued*) at ϕ deg. The line target must then be seen at θ deg from the frontal plane because only in this orientation does it project the identical pattern of binocular disparities as the actual frontal-plane target straight ahead of the observer. Ebenholtz and Paap (1976) showed these deductions to be largely fulfilled when ocular direction was biased either by the aftereffects of wearing a prism or sustained asymmetry of gaze.

all else is constant, a shift in apparent direction of gaze or in the gaze direction registered by the oculomotor system produces a corresponding rotation in the AFP. The geometry requires the rotation to be twice the displacement angle and when the displacement is produced by wedge prisms (see Chapter 2) the expectation was nearly perfectly corroborated (Ebenholtz & Paap, 1976). Additional tests of the model based on aftereffects of prism exposure or of sustained leftward or rightward ocular deviation also supported the model. Although orientation changes were as predicted, direct measures of displacement based on a pointing response generally were low and did not correlate well with the measured shifts in orientation. Nevertheless, because the operations chosen by Ebenholtz and Paap were known on previous grounds to bias oculomotor direction, there is little doubt of the influence of this factor on AFP.

Perceptual Effects of Biased Vertical Eye-Position Information

a) Apparent horizon and the pitch box. Just as in the case of horizontal eye position, a bias in the resting level of the eye-centering response in the vertical direction also should produce shifts in apparent direction of a visual target. One special class of these effects refers to the apparent horizon because this is determined typically by having subjects direct the gaze so as to place the lines of sight parallel with the ground plane, as though looking straight ahead at the nearly infinitely distant horizon. When, in complete darkness, subjects adjust a spot of light to appear at the height of the horizon it typically is set 2 to 3 deg lower than objective eye level (Stoper & Cohen, 1986; Matin & Fox, 1989). Furthermore, when also in total darkness, simply attempting to look straight ahead or to place their lines of sight horizontal with the ground plane, subjects direct their gaze from 3 to 6 deg below objective horizontal (Cohen, Ebenholtz, & Linder, 1995). Thus the misdirected eye position would seem to be responsible for the lowered apparent horizon, or, by inference, the judgment that a target actually at true eye level appears inordinately elevated.

One way to influence the vertical direction of gaze is by way of a pitch room as shown in Fig. 4.07. When the room, within which the subject is enclosed, is pitched, two related effects occur. First, a spot of light adjusted to appear at eye level or at the horizon actually is moved in the same direction as the front of the room. For example, upward target displacement occurs with upward pitches of the front wall of the room where the top of the wall is pitched toward

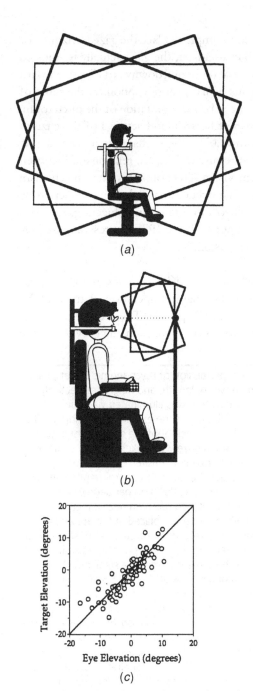

(a)

(b)

(c)

Figure 4.07. (a) *Pitchroom. (From Cohen, Ebenholtz, & Linder, 1995.* Effects of optical pitch on oculomotor control and the perception of target elevation, *Perception and Psychophysics, 57,* 433–440. (*Continued on next page*)

the observer. Second, with instructions simply to place the gaze horizontal or to look straight ahead, with no target to adjust, the gaze actually is adjusted upward with upward pitch of the front of the room, downward with downward room pitch (Cohen et al., 1995). Thus ocular proprioception, or the sense of where one is gazing, is heavily influenced by the orientation of the pitch room. Are settings of a target to the apparent horizon and settings of the gaze to apparent horizontal correlated? Figure 4.07(c) shows that they are very closely related phenomena supporting the premise of a causal relationship between ocular proprioception and apparent visual direction. This same mechanism probably also is at work in illusions that occur in natural settings where truly horizontal surfaces appear to be pitched and rotated and objects appear with erroneous heights. These "mystery spots" (Ross, 1975) represent extremely compelling versions of the pitchbox illusion.

b) Apparent visual direction and the doll reflex. Yet another way to bias eye position information is by way of the natural operation of the otolith–ocular reflex known as the *doll reflex* (see Chapter 3). These eye movements occur in the vertical plane and are driven by changes in the gravito-inertial

---

Figure 4.07. *(Continued)* Reprinted with permission of Psychonomic Society, Inc.) The room measured 1.2 × 2.8 × 1.7 m in width, length, and height, respectively. Walls were covered with a dense grid pattern that was illuminated via indirect lighting. Black cloth covered the floor and all view of the exterior surroundings. The room pivoted about an axis approximately at the height of the subjects interaural axis, and measurements of apparent horizontal gaze were taken at pitch angles of 0-, 10-, and 20-deg above and below true horizontal. Infrared eyemovement recordings revealed a significant linear relationship between the room pitch and eye position, with a slope of from 0.3 to 0.4 deg of eye elevation per degree of room pitch. (b) *Pitchbox.* Does the influence of the pitched room on gaze direction correlate with the effect of room pitch when a target is placed at the apparent horizon? To answer this question, Cohen et al. recorded both the eye position when the gaze was set to be apparently horizontal, and the target position when it was set to appear at the level of the horizon. For these measurements a small pitchbox placed close to the subject was used with dimensions 30.5 × 30.5 × 45.7 cm in depth, width, and height, respectively. A small target LED that moved vertically along the far wall and electroluminescent strips about 6.3 mm wide that ran along the intersections of the front and side walls were visible. (c) *Relationship Between Target and Eye Elevation When Set at the Apparent Horizontal and at Level Gaze, Respectively.* The two measurements taken on *separate* sessions were compared. There were 16 subjects and 5 pitchbox angles; therefore, 80 paired observations are represented. The coefficient of determination r^2 indicated that 75 percent of the variance in target elevation was accounted for by the variance in eye elevation.

forces acting on the otolith organs. Although the reflexive eye movements themselves have no conscious correlate, fixating a target after the whole body is placed at some given pitch angle does lead to the perception of illusory target height. In the 90-deg range of body pitch from upright to supine, the eyes are driven downward with backward pitch (Ebenholtz & Shebilske, 1975; Citek & Ebenholtz, 1996). As a consequence, a point-target chosen as appearing straight ahead actually was lower than the one chosen with the body upright (Ebenholtz & Shebilske, 1975), and a target fixed in space and seen as straight ahead at the upright appeared to be too high when viewed from the supine position. Some elements of the measurement technique are represented in Fig. 4.08. The apparent elevation probably reflects the voluntary innervation necessary to raise ocular gaze in order to maintain fixation against the downward pull of the reflex. As a rule, perceptual illusions result whenever there is a balancing of voluntary against reflexive innervation because only the former tend to provide salient, conscious representation.

The doll reflex also can be biased by adaptation. Shebilske and Karmiohl (1978) had their subjects maintain fixation in the dark on a small luminous target after they were pitched back by 40 deg. The target dot had been placed at the subjective straight-ahead position selected by the subject when at the upright orientation, therefore sustained fixation required a voluntary effort to counteract the downward pull of the doll reflex. After a 5-min fixation period, subjects showed about 3 deg of upward shift in their choice of a target that appeared to be straight ahead. The same shift occurred whether subjects were pitched or upright during the pre- and posttesting phases. Shebilske and Karmiohl thus showed that the doll reflex is capable of adaptation that is independent of body pitch, as if some constant degree of innervation had been added to the muscle tonus controlling the straight-ahead eye position.

c) Induced vertical phoria and apparent height. It has been known for some time that a base-up prism over one eye (see Fig. 4.10b) and a base-down prism or no prism at all over the other eye, if of sufficiently low power, induces each eye to move to eliminate the visually displaced images (Ellerbrock & Fry, 1941; Ogle & Prangen, 1953). Interestingly, when prisms were removed after varying time periods of sustained fusion, evidence was found of an induced vertical phoria (IVP), such that each eye pointed in the direction of the previous image displacement (Ogle & Prangen, 1953). In other words, the lines of sight of each eye pointed in a vertical plane but in opposite directions. Will the apparent height of a target vary accordingly when viewed separately by each eye? When the vertical resting level of an eye has been reset, then in order to fixate a

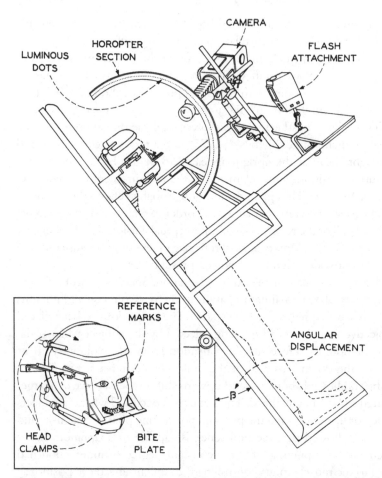

Figure 4.08. *Subjective Measure of the Doll Reflex.* (From Ebenholtz & Shebilske, 1975, The doll reflex: Ocular counterrolling with head–body tilt in the median plane. *Vision Research, 15*, 713–717, Copyright 1975. Reprinted with permission from Elsevier Science.) Subjects selected from a vertical array of luminous dots, the one that appeared to be straight ahead and with gaze perpendicular to the subjects' frontal plane. Selection was made with the subject upright and at 10-deg steps in backward pitch up to 90 deg. Results showed that as backward body pitch increased, subjects chose a target as straight ahead that actually was lower and lower on the horopter. The camera shown in the figure was used to verify that actual eye position also changed correspondingly. Despite the downward rotation of the globe, subjects believed they were pointing the gaze in the same direction relative to the head at each body-pitch angle. Thus, as in the case of ocular counterrolling and even nystagmus (see Chapter 3), there appears to be no conscious correlate of reflexive eye movements.

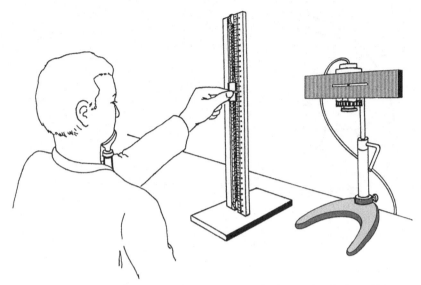

Figure 4.09. *Manual Measure of Apparent Height.* (From Ebenholtz, 1978, After-effects of sustained vertical divergence: Induced vertical phoria, and illusory target height. *Perception, 7,* 305–314. Reprinted with permission of Pion, London.) Before and after the induced vertical phoria, subjects matched the tactually perceived ridge on a small vertical cylinder to the apparent height of a thin luminous line. Viewing was done monocularly with each eye tested in succession and in complete darkness, except for the horizontal line.

The vertical phoria was induced by having the subject view the luminous line binocularly with a variable power base-up or base-down prism over one eye. Prism power was increased to the maximum consistent with singleness of vision, and fusion was then maintained for 6 to 8 min. This procedure produced an induced vertical divergence of about 2.00 *D*, or 1.15 deg.

target at some standard height, it seems reasonable to conclude that the added innervation necessary to move from the biased resting level to fixate the target should signal a change in target elevation. For example if an eye rests slightly downward reflecting its IVP position, then the "effort" to fixate should yield the experience of increased elevation, and conversely for an eye that is made to rest in an upward gaze. This hypothesis was tested by having subjects adjust a moveable tactile object to match the apparent height of a monocularly viewed target, as shown in Fig. 4.09. The direction of differences in height settings between the two eyes was found to be dependent on the direction of bias in accordance with expectations (Ebenholtz, 1978). Thus, there is evidence that to some extent each eye is capable of being independently biased, and for the independent expression of these effects on apparent height. Paap and Ebenholtz (1977) also found eye-specific adaptation after subjects sustained convergence

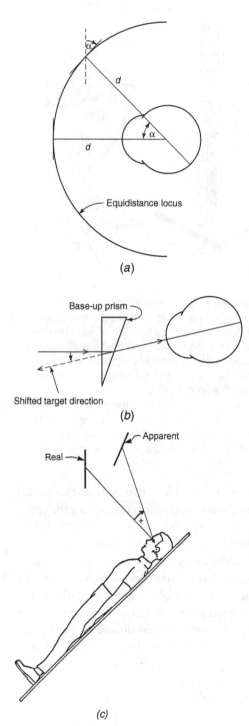

(a)

(b)

(c)

Figure 4.10. (Continued on next page)

for 10 min on a near target at about 12 cm. In this case the shift in apparent straight ahead occurred in the horizontal plane, with each eye signalling a change in direction opposite to that of the other eye. Both sets of results are consistent with the hypothesis that during monocular viewing only the open eye determines the apparent direction of visible objects (Erkelens, 2000). There is evidence to the contrary, however, (see Figs. 4.01(d) and 4.04); therefore, additional research is needed to establish just what constraints to Hering's laws of visual direction may be needed.

d) Apparent vertical orientation in the pitch plane. Changes in apparent target elevation that accompany either the doll reflex or induced vertical phoria are a relatively direct result of reflexive innervation of upward- and downward-acting eye muscle, but the effect of a vertical bias on the apparent vertical orientation of a line is somewhat more complex. In fact it requires a model very similar to that of the Vieth–Müller circle (see Fig. 4.05) used to explain change in the apparent frontal-plane orientation with change in horizontal ocular direction. The model, an equidistance locus, is represented in Fig. 4.10(a). It presumes that with visual input constant, a vertical target seen with increased elevation also will be seen as changing its apparent orientation in a one-to-one relation, as if the target was simply slid along tangent to the equidistance surface.

Ebenholtz and Paap (1976) found moderate support for these expectations: after a 30-min exposure to base-up or base-down prisms, 24 of 32 subjects

---

Figure 4.10. *(Continued) Model of Change in Apparent Vertical with Ocular Elevation.* (a) An eye is represented from the side, viewing a vertical target at a fixed distance *d* from the nodal point. When innervation to the ocular muscles involved in elevation is biased, as a result of an aftereffect of sustained viewing through base-up prisms (b) or during backward pitch (c), a target objectively straight ahead and vertical is registered as higher than otherwise. The example in (c) (modified from Ebenholtz, 1977a, The constancies in object orientation: An algorithm processing approach. In W. Epstein (Ed.), *Stability and Constancy in Visual Perception: Mechanisms and Process.* Copyright 1977, John Wiley & Sons. Reprinted by permission of John Wiley & Sons, Inc.) assumes a backward body pitch such that a downward acting ocular reflex occurs. Because maintaining fixation requires innervation to counter the reflex, the net effect is as if the gaze was directed upward, through an angle α in (a) and (c). The model presumes the target will be seen as though it is rotated upward staying at a fixed angle to the line of sight, and in so doing, changing its apparent orientation, increasingly top toward the observer. By remaining at a fixed angle to the line of sight, the target projects the same retinal image at any gaze angle along the equidistance locus. The shift in apparent orientation angle is equal to the displacement angle α.

demonstrated the predicted shifts in line orientation. Support also has been found for the effects of the doll reflex on apparent vertical orientation (Ebenholtz, 1970), as represented in Figs. 4.10(c) and 4.11. When subjects who were pitched backward adjusted a luminous line to appear vertical (within the median plane), they required the top of the line to be rotated top away, as expected from the model. However, beyond about 45 deg of backward body pitch, a reverse error occurred. The entire function relating deviation of line from true vertical to body pitch is shown in Fig. 4.11. Because the line cannot be set to gravitational vertical without registration of the magnitude of body pitch, it may be concluded that more factors than the doll reflex must be involved in judgments of the apparent vertical. It would appear that the doll reflex is expressed at small body tilts whereas increasing underregistration of body pitch operates perhaps throughout the entire range of body pitch tested.

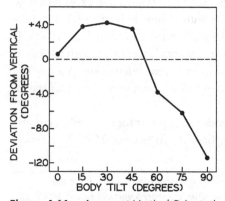

Figure 4.11. *Apparent Vertical Orientation with Backward Body Pitch.* (From Ebenholtz, 1970, Perception of the vertical with body tilt in the median plane. *Journal of Experimental Psychology, 83,* 1–6. Copyright © 1970 by the American Psychological Association. Reprinted with permission.) Subjects viewed a thin luminous line [as represented in Fig. 4.10(c)] and had adjustments made to its orientation until it appeared to be vertical. Positive errors indicate that the line was rotated beyond true vertical top away from the subject, whereas negative errors represent the rotation of the line, top toward the observer. The positive errors at small body pitches are consistent with the model of the influence of the downward acting doll reflex as shown in Fig. 4.10(a and c). The negative errors at larger body pitches indicated that a truly vertical line appeared rotated, top away, and therefore required adjustment, top toward the observer, in order to appear upright. If the nervous system were to register *less* body pitch than actual, the negative errors could be accounted for because there would be less need to compensate for body pitch by rotating the line out of parallelarity with the body. There is, however, no direct evidence for this hypothesis. In fact, to the contrary, estimates of body pitch in darkness (Ebenholtz, 1970), suggest increasing apparent pitch well beyond actual. It is possible that the proper measure of body pitch has yet to be found.

e) Apparent vertical and horizontal orientation in the frontal plane. A very similar function to that of Fig. 4.11 also characterizes the error in setting a line to the apparent vertical during head and body tilt to one side (i.e., as a result of roll tilt). Up to about 40 deg of lateral body tilt, the target line, usually luminous in a dark room, must be rotated in a direction opposite that of the body in order to appear gravitationally vertical. This has been termed the *E-effect* (Müller, 1916).

With increasing body tilt the line must be set increasingly in the same direction as body tilt toward parallelarity with the longitudinal body axis. As Fig. 4.11 shows, the errors on the right side of the curve, known as the *Aubert* or *A-effect* (Müller, 1916), are much larger and opposite in direction to the E-effect.

Because of the small magnitude of the E-effect that approximates the size of either the doll reflex or ocular counterrolling, it has seemed plausible that these otolith–ocular reflexes may actually underlie both types of E-effect. A model to account for the E-effect derived from pitch-plane tilts and based on the doll reflex has been presented previously. An analogous account of E-effects for lateral roll tilts also has been developed based on the counterrolling reflex (Ebenholtz, 1970), but it is only more recently that careful measurements of torsion and ocular counterrolling have been made and compared with measures of apparent vertical or horizontal in the same experimental context (Wade & Curthoys, 1997; Nakayama & Balliet, 1977).

The underlying logic is that ocular counterrolling and torsion, being re-flexive, have no conscious correlates nor leave any record of torsion in the nervous system. Therefore, a pattern of retinal stimulation falling on a torted eye is treated as though the eye had not undergone torsion or counterrolling at all. It is as though the system responsible for perceptual processing assumes the eye to be in its default torsional position. This target would appear upright when its image would fall on or near the vertical retinal meridian, regardless of the actual torsional position of the globe in the skull. In contrast, if torsional position were registered in the nervous system, with or without conscious correlates, then this information could be used to reposition the locus on the retina that, when stimulated, would give rise to the apparent vertical. In an ideal sense, in order to preserve veridical perception this locus would change exactly by the amount of torsion. For example, if the vertical retinal merid-ian of an upright observer torted by 5 deg clockwise, the retinal locus for the apparent vertical would have to be shifted 5 deg counterclockwise for veridical perception of the vertical of space.

To test these alternatives Nakayama and Balliet (1977) stimulated a torsional response by having the subject assume an oblique or tertiary position of gaze (e.g., up and to the right, down and to the left). Such eye positions follow List-ing's law (Howard, 1982, pp. 180–186), which requires the eye to rotate around a single oblique axis in *Listing's plane*, a plane that approximately parallels the frontal plane running through the center of ocular rotation. Because tertiary gaze directions arise from ocular rotations around oblique axes, they typically produce clockwise torsions when looking up and to the right or down and toward the left, whereas counterclockwise torsions accompany eye positions up and toward the left, and down toward the right. Accordingly, Nakayama and Balliet (1977) recorded both the ocular torsional position as well as the position of an electroluminescent line when judged to be vertical while the eye was in a tertiary position of gaze. The results, shown in Fig. 4.12, indicate that

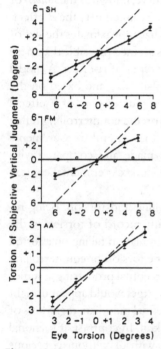

Figure 4.12. *Line Settings to Apparent Vertical as a Function of Ocular Torsion.* (Reprinted from, Nakayama & Balliet, 1977, Listing's law, eye position sense, and perception of the vertical. *Vision Research, 17*, 453–457, Copyright 1977. Reprinted with permission from *Elsevier* Science.) If there were no compensation for ocular torsion, the line settings to apparent vertical would fall on the dashed line. They would simply track the degree of ocular torsion, always stimulating the same retinocortical locus. If torsion were perfectly compensated for, all settings would fall along the horizontal line as do those of subject FM, open circles, which were taken in a fully lighted room. Note that although the authors found the deviation from the torsion prediction, which represents the extent of compensation, to increase quite linearly with increasing degrees of ocular torsion, substantial differences in the magnitude of compensation exist among individual subjects.

the line settings characteristically were rotated less than the degree of ocular torsion. Therefore, the line settings did not simply track the ocular torsion (dashed lines) but neither were they veridical, as represented by the horizontal line. The authors suggested that the ocular torsional signal, although sufficient to provide a degree of compensation, was not sufficient to allow a fully veridical percept. Nevertheless, the interesting implication remains that an ocular torsion signal was available to modify the retinocortical locus of stimulation that would otherwise represent the apparent vertical of space. It also seems to be the case that this use of the torsion signal can occur quite independently of one's awareness of the torsion itself. It is worth noting that a similar division of signal use occurs with optokinetic afternystagmus (OKAN), the continuation in darkness of optokinetic nystagmus (OKN) after the withdrawal of OKN stimulation. In this case, although one generally is not aware of the nystagmus eyemovements, one can nevertheless point properly in the direction of a flash of light that is timed to occur during the nystagmus movement (Bedel, Klopfenstein, & Yuan, 1989). It follows that a signal conveying eye position is available to the manual system, even though it does not provide for awareness of eye position during OKAN. It is not known whether VOR follows a similar rule.

Two other methods for producing ocular torsion were employed by Wade and Curthoys (1997) to further test for the relationship between ocular torsional position and apparent line orientation. In one case, subjects were rotated at the end of the arm of a centrifuge, 1.0 m from its center. Rotation velocities were chosen to produce roll-tilt equivalents of 10-, 20-, 30-, and 40-deg corresponding to the angle θ in Fig. 4.13. The resultant of the centrifugal and g-force levels, as shown in Fig. 4.13, acts to stimulate the otolith organs in the same direction as if the subject actually were tilted in a 1-g field, but with a somewhat greater force on the otoliths (Howard, 1982, pp. 438–439). Otolith–ocular reflexes are thus triggered as long as the subject continues to rotate. Subjects set a luminous line to apparent horizontal, had infra-red measures taken of the iris, from which ocular torsion was measured, and in darkness rotated a bar, grasped with two hands, to the apparent horizontal. The authors assumed that the luminous line setting was the result of two additive components, one of which was ocular torsion, the other being a nonvisual somatosensory indication of the gravitational horizontal that might itself reflect some input from the sense of postural roll tilt. Certainly, if the magnitude of body roll is not properly registered, then the visual retinal locus that would register as apparent horizontal could not be properly chosen. Imagine lying on your side but perceiving that you were upright while setting a line to be apparently horizontal. Under this unlikely condition and setting considerations of torsion aside, temporarily, a gravitationally horizontal line would fall near the

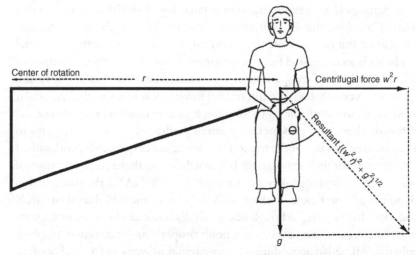

Figure 4.13. *The Approximate Equivalence between Centrifuge-Induced Tilt and Actual Tilt in a 1-g Field.* (From Howard, 1982, *Human Visual Orientation*, p. 437. Copyright 1982, John Wiley & Sons, Limited. Reproduced with permission.) Howard (1982) points out that in a centrifuge the resultant force always is greater than 1.0 g. This follows from the analysis of forces generated at distance *r* from the center of centrifuge rotation at a radian velocity ω. The centrifugal force equals $\omega^2 r$, which may be divided by 9.8 m s^{-2} to change it into units of g-force. The resultant is the hypotenuse of the right triangle that equals $\sqrt{(\omega^2 r)^2 + g^2}$. The angle at which the force acts may be gotten from the tangent relation $\tan \theta = \omega^2 r/g$. In Fig. 4.12, the subject would feel upright when aligned with the resultant; therefore it may be inferred that the subject experiences a body tilt toward the left side.

At a rotation speed of 164 deg sec^{-1}, one of the speeds used by Wade and Curthoys (1997), $\omega = 2.86$ rad s^{-1} and with a 1 m arm, $\omega^2 r = 8.192$ or .836 in units of g. Accordingly, the resultant force equals $\sqrt{(.836)^2 + 1.0}$, or 1.303 g units and because $\tan \theta = .836$, $\theta = 39.9$ deg. If a subject actually were tilted at 39.9 deg, then the force needed to keep from falling would be a lateral force F_L equal to $\sin \theta$ or .641 g. In the centrifuge, the lateral force was the centrifugal force $\omega^2 r$, which equaled .836 g, an increase of more than 30 percent. Thus, as Howard (1982) suggested, the centrifuge paradigm may enhance the sense of body posture in comparison with that measured in a 1.0 g field.

vertical retinal meridian, but perceiving yourself as upright, it would appear to be vertical and would therefore be rejected. Thus, postural inputs are critical to the process and these presumably were represented in the somatosensory measure of the gravitational horizontal. Based on the algebraic model:

visual line settings = torsion + somatosensory settings

therefore

torsion = (visual line − somatosensory) settings.

The results were remarkably consistent in demonstrating high correlations, ranging from 0.85 to 0.99, between the difference (visual — somatosensory), representing torsion derived from the two perceptual measures and the objectively measured torsion. Because the actual torsion was in nearly perfect correspondence with the derived difference score, there was no evidence for a torsional compensation process. Put another way, the residual remaining after subtracting the somatosensory component from the visual target settings equaled the actual torsional response with little else to account for.

Yet another technique to induce ocular torsion, but without disturbing postural orientation, is to accelerate the subject in yaw rotation around a vertical axis (Smith, Curthoys, & Moore, 1995). During the acceleration phase, rightward rotation produced counterclockwise torsion as viewed from behind the subject, whereas torsion reversed for leftward rotation. For measures taken during or shortly after the angular acceleration, Wade and Curthoys (1997) found the visual settings of apparent horizontal to be nearly completely accounted for in terms of the torsion response. Thus, both studies found no evidence for a compensatory process that might take account of the torsion and adjust the retinocortical locus for settings to the apparent vertical and horizontal. Of course, it is possible that the evidence for compensation found by Nakayama and Balliet (1977) may be related to the voluntary nature of the eyemovements their subjects used to place the eyes in tertiary positions of gaze. The question calls for direct comparisons between voluntary and reflexive eyemovement paradigms.

Perceptual Effects of Vergence and Accommodation

Direct Affect on Distance Perception

The notion that convergence and accommodation signal apparent distance was first put forth in the modern era by Descartes (1664/1972). For Descartes, the relation between changes in lens curvature and ocular vergence, the ensuing brain stimulation (which he thought to be of the pineal gland) and sensed distance was an innate one. Subsequent philosophers, such as Berkeley (1709/1948), however, believed the connection among ocular vergence, accommodative blur, and distance to be arbitrary and established by individual experience. At the end of the twentieth century, there existed dozens of experiments on the nature of the relationship between convergence and accommodation on the one hand and sensed distance and size on the other (Boring, 1942; Leibowitz & Moore, 1966; Kaufman, 1974; Foley, 1980). Although it still is not

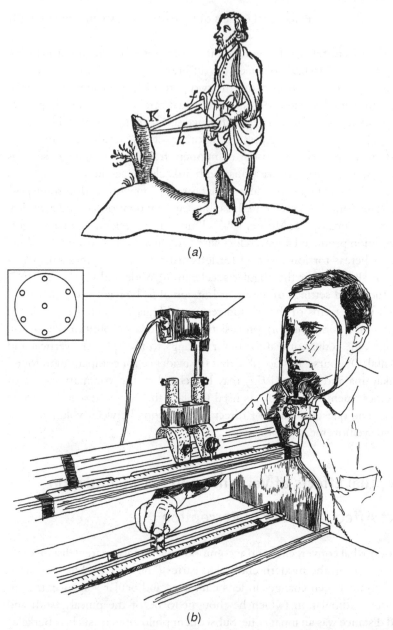

(a)

(b)

Figure 4.14. (a) *Distance Perception Via "Natural Geometry."* (From Descartes, 1664/1972, *Traite de L'Homme)*. There appears to be no research on either blind or sighted subjects to verify or refute Descartes' claim that distance may be perceived by manual triangulation, as the sketch in (a) implies. Descartes' application of this idea to the vergence system, however, has much more support. The trigonometry is shown in Fig. 3.05, where it is demonstrated that the separation between the eyes or interocular axis *a* and the ocular vergence angles *(Continued on next page)*

known whether a "natural geometry" underlies vergence-mediated distance perception, as Descartes had proposed, some evidence favors this view. For example, when the convergence angle of 4- to 8-month-old infants was altered by prisms placed in front of their eyes, the babies did misreach in the expected direction (Hofsten, 1977). Descartes may have been correct, or if experience was required, as Berkeley believed, it must occur quite early in life.

Descartes believed that a blind person shown in Fig. 4.14(a) could sense the distance to an object or surface through the operation of "natural geometry," where the distance between the hands and the angle at which the hands were rotated convey sufficient information to infer distance. The reasoning is analogous to the establishment of the complete identity of two triangles in Euclidean geometry by demonstrating equality in only one side and two adjacent angles.

Figure 3.05 demonstrates the application of geometry and trigonometry to the vergence system, and Fig. 4.14(b) demonstrates a method for measuring distance perception while converging on a target. The method is quite sensitive to changes in the resting level of vergence such as those brought about by sustained binocular fixation at a near or far target (see Chapter 3). For example, after working at a near distance for about 6 to 8 min, the resting level of vergence, as well as accommodation, shift inward. Because these resting levels represent a type of physiologic null point or fulcrum (Tyrrell & Leibowitz, 1990; Ebenholtz, 1992) about which accommodative and vergence effort varies, any shift in the resting level entails a shift in effort needed to focus or verge. In general, inward shifts require less effort than previously to verge and focus, whereas outward shifts require increased effort. The "effort" is, of course, a subjective state that may not always be experienced, but at a physiologic level it corresponds to a change in innervation frequency or impulses per unit time, sent to the ciliary or extraocular muscles. It is the registration of this change

---

Figure 4.14. (*Continued*) $\gamma/2$ suffice to deduce distance d, namely, $d = a/2\tan(\gamma/2)$. (b) (From Ebenholtz, 1981, Hysteresis effects in the vergence control system: Perceptual implications. In D. F. Fisher, R. A. Monty, & J. W. Senders (Eds.), *Eye Movements: Cognition and Visual Perception*, Chap. 11.2. Reprinted with permission of author.) *A Manual Measure of Target Distance*. Typically, the measure is taken in darkness so that only the target is visible, but not the arm or pointing finger. In general, subjects are capable of reliably placing the finger beneath the apparent location of the visible target. Unpublished studies of the author have shown that, in symmetric fashion, with pointing finger fixed in place, the target or convergence distance can be adjusted reliably to match the location of the fingertip. Thus both the visual and manual systems are capable of using distance information originating in the opposite channel.

in the cortex that is probably correlated with a change in distance perception. In the apparatus shown in Fig. 4.14(b), the result of sustained vergence on a near target is to place the pointing finger beyond its original prefixation position, thereby indicating a greater distance perception for the target. As might be expected, nearer pointing positions result from sustained far fixation (Ebenholtz & Wolfson, 1975; Ebenholtz, 1981). It is clear that a systematic bias induced in the vergence resting level causes a systematic shift in apparent target distance (Ebenholtz & Fisher 1982; Owens & Leibowitz, 1983). Because other information that may otherwise be thought to produce changes in target distance has been carefully excluded by the adaptation paradigm, and because a bias in the vergence resting level causes shifts in apparent distance, then it may be concluded that vergence is indeed a contributor to sensed distance. Is this also true of accommodation?

Although many studies have implicated accommodation in near-distance perception, few actually have measured both the accommodative response and perceived distance in the same context. This was accomplished by Fisher and Ciuffreda (1989), who followed up a lead found in a prior study (Ebenholtz & Fisher, 1982) in which subjects walked through a hallway for 15 min while viewing monocularly through a 5.5 diopter negative lens (see Chapter 2). Such a lens changes the light vergence so that the farthest object in the environment provides light rays as though it were at only 18 cm. Although accommodation was not measured directly in the Ebenholtz and Fisher study because subjects maintained clear vision it was inferred that they responded with increased accommodation to the high accommodative demand imposed by the lens. Using the pointing apparatus shown in Fig. 4.14(b) with monocular testing before and after a sustained focusing task, a slight but significant *increase* in pointing distance was found after the monocular exposure period. This outcome is consistent with an *inward* shift in accommodative resting level, thereby requiring reduced effort to focus on targets. If one thinks of the resting level as a fulcrum or point of balance between near and far focusing, then an inward shift of this point would place near targets dioptrically closer to the resting level than previously and hence less effort to accommodate would be required. Likewise, targets on the far side of the resting level would be dioptrically further from the fulcrum and therefore more *relaxation* effort would be required. In both cases, a target more distant than previously shown would be signaled, and in fact Fisher and Ciuffreda (1989) verified these deductions by showing that inward shifts in accommodative resting level were accompanied by corresponding outward shifts in pointing distance.

An interesting result of this study occurred when subjects made distance judgments by pointing toward a low spatial-frequency accommodative target, such as a luminous disk with indistinct borders. Under these conditions the accommodation would not be engaged and the focusing mechanism fell to its resting level. In their study, however, there were two separate measures of resting level: One before and the other after the 10 min sustained focusing period. By comparing the two, Fisher and Ciuffreda found that regardless of the significant shift in resting levels induced by the 10 min task, the pointing response to the low frequency test target remained relatively unchanged. Thus, although the authors did not draw this conclusion, their data suggest that the accommodation resting focus, no matter what its level, corresponds to a fixed apparent distance. This is consistent with the concept of resting level as a reference magnitude against which changes, both in accommodative effort and apparent distance, are measured.

Indirect Affect on Apparent Size and Apparent Depth

The studies described previously demonstrate an affect of oculomotor bias directly on the perception of distance. However, it is to be expected that those perceptual qualities that *depend* on distance also should be influenced by oculomotor adjustments. Accordingly, we turn to the perception of size and depth for such instances.

The visual perception of size is mediated by the size of the region of stimulated photoreceptors on the retina, in short by the size of the retinal image. However, because the linear size of the image varies inversely with the distance to the corresponding external object, image size alone is not sufficient to specify an unchanging perceived size. In the mid-1600s, Descartes summarized the problem that has come to be labeled as "size constancy" by observing that "while the images may be, for example, one hundred times larger [in area] when the objects are quite close to us than when they are ten times farther away, they do not make us see the objects as one hundred times larger because of this, but as almost equal in size, at least if their distance does not deceive us" (Descartes, 1637/1972, p. 107). Because although images of distant objects shrink their perceived size stays relatively constant, a good test of the role of distance information in this process would ensue if on keeping image size fixed, variations in the distance-signaling mechanisms were to produce corresponding variations in perceived size. A seemingly ideal way to establish a fixed retinal image is to fixate the flashbulb of a camera that produces an image that outlasts the flash itself, hence the term *afterimage*. Viewing the afterimage against the palm of one's hand produces the appearance of a rather small blob, whereas seeing it on a wall across the room elicits the appearance of a distinctly

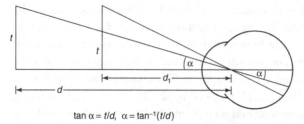

$$\tan \alpha = t/d, \quad \alpha = \tan^{-1}(t/d)$$

Figure 4.15. *Emmert's Law and Size Constancy.* A target *t* at distance *d* produces an angle *α* at the retina that would increase if the target were moved to a shorter distance d_1. The formula shows that this angle would remain fixed at the retina so long as *t* and *d* were kept in the same proportions. For example, if distance doubled, then if the target size also doubled, the retinal angle would remain constant; conversely, if the retinal angle were fixed, then target size must be made proportional to distance. These physical relationships may have provided the underlying model for Emmert's (1881) law, which states that with a fixed size of afterimage, its perceived size will be proportional to the perceived distance at which it is located.

Emmert's law may be understood as contributing to size constancy by balancing two opposing factors. For one, the larger *image* produced by *t* at the nearer distance d_1 would contribute to an increase in apparent size. However, this would be opposed by the smaller *apparent size* required by Emmert's law to be proportional to the shorter apparent distance. Approximate size constancy may be understood as the result of the summation of these two processes.

Emmert's law also plays a critical role in explanations of the moon illusion, where the horizon moon may be seen as having nearly twice the diameter of the moon at the zenith (Kaufman & Rock, 1962). The application of Emmert's law presumes that distance cues, available from the terrain when viewing the horizon moon, signal a greater distance than the cues present when viewing the moon elevated in the sky. Because the angular size of the moon at the retina is essentially constant, at 0.52 deg, Emmert's law requires the moon on the horizon to be processed as though at a greater distance and, therefore, to be seen as larger (Kaufman & Kaufman, 2000).

larger blob. These observations conform to the classic formulation of Emmert (1881) to the effect that the apparent size of an afterimage is proportional to its apparent distance. Such a quantitative statement, now known as Emmert's law (demonstrated in Fig. 4.15) is, of course, a valuable enhancement of Descartes' observation and illustrative of the rule that empiricism serves as a significant bridge between philosophy and science.

Unfortunately, the case for perceived distance is not quite made with these observations because in relation to the large image of the near palm the afterimage may be small whereas in relation to the image of the distant wall, the afterimage may be large. Therefore, relative size may be mediating the changes in apparent size of the afterimage. A relative-size illusion is shown in Fig. 4.16.

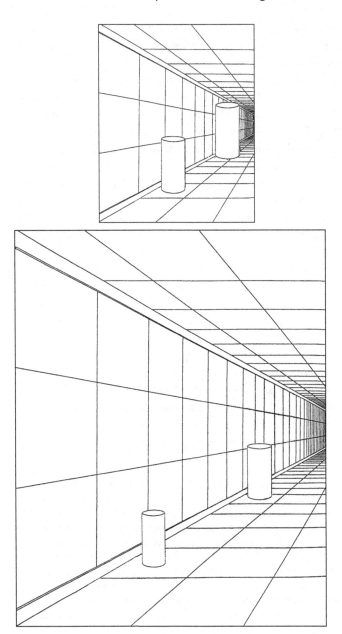

Figure 4.16. *Relational Factors in Size Perception.* Within each hallway figure, the farther cylinder appears larger than the nearer one, demonstrating the effect of apparent distance on perceived size (Gibson, 1950). The cylinders in the upper and lower figures, however, have different relationships with their respective hallways. Because apparent distance is a known factor in size (*Continued on next page*)

Since Emmert first formulated his law, a large number of studies have attempted to verify it, as the study of size perception shows. One of the most elegant researches in this genre in which image size is held fixed and distance cues are manipulated while apparent size is measured is that of Leibowitz and Moore (1966). They manipulated distance information by systematic variation in convergence and accommodation while keeping image size of a real target fixed at 1.0 deg of arc. Figure 4.17 represents the outcome of this procedure for a test target at an actual distance of 50.0 cm from the observer. It is clear that convergence and accommodation are capable of systematically altering the apparent size of an object whose image size is fixed at the retina. It seems likely that the same process operates to modulate size perception even when image size is *altered* (e.g., by changing the observation distance; see Fig. 4.14). Because the image size of an object increases at near distances and decreases at far distances, a useful and functional rule for the visual system would be to undo the potential perceptual effects of these changes. As Fig. 4.17 shows, this is evidenced in the result that for a given image size, the apparent size is diminished at near optical distances, where the image is large, and enhanced at far distances, where the image is small, thereby achieving a degree of constancy in apparent size.

Just as the perceptual system can use convergence to scale image size, so too can it use the same oculomotor cue to scale binocular disparity (see Fig. 3.08) to produce a constancy of apparent depth (Wallach & Zuckerman, 1963). In fact, if disparity is held fixed while convergence is relaxed, apparent depth will grow in a manner analogous to Emmert's law, although because of geometric considerations, apparent depth is expected to grow more rapidly than size, as

---

Figure 4.16. (*Continued*) perception, the two cylinders within each figure were placed so as to approximate equal apparent distances with their counterparts in the other figure, relative to the picture plane. The upper cylinders fill up more picture space than do the lower ones. This may be the reason why the upper pair appears somewhat larger than the lower one, because all cylinders actually have identical dimensions. The effect could be purely geometric (Rock & Ebenholtz, 1959), according to which, with apparent distance held constant, objects appear to be unequal in size when they occupy disproportionate portions of their respective local backgrounds. Alternatively, the effect may entail cognitive processing concerning the normative sizes of hallways, reflecting a syllogistic process that is presumed to influence perception (Rock, 1983). For example:

1. The two patterns are identified as hallways and as such they have typical hallway dimensions.
2. The upper cylinders take up more space than do the lower ones.
3. therefore . . .

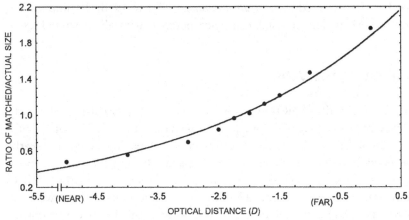

Figure 4.17. *Ratio of Matched to Actual Size as a Function of Optical Distance.* (Based on data of Leibowitz & Moore, 1966). The target, a white equilateral triangle 0.87 cm in height, was viewed at a 50 cm distance against a black background and produced a retinal image of precisely 1 deg of arc in height. Subjects signaled its apparent height by adjusting the height of a comparison triangle at 2 m distance so that the two appeared equal. The target was viewed through combinations of prisms and lenses creating a range of optical distances from infinity to −5 D or its equivalent of 20 cm. The ratio of matched to actual size steadily increased as optical distance increased; namely, the target appeared almost 50 percent smaller than actual size at near distances, and nearly twice as large at far distances, even though retinal size was fixed. Because no other images were present along with the target, it is unlikely that Emmert's law of the apparent size of after images could be an artifact of *relative* image sizes.

It is important to consider how the combined effects of convergence and accommodation on apparent size may contribute to size constancy, over the range of optical distances shown in the figure. The nonlinear aspect of the function shows that a 1 *D* shift in optical distance produces different changes in apparent size, depending on whether the shift was at near or far distances. The 1.0*D* shift from −1 to 0 diopters caused an increase of 34 percent in the size ratio, but of only 17 percent after changing from −5 to −4*D*. It is appropriate that there be less change in apparent size at near distances because the 1.0 *D* shift at far covers the distance from 1 m to infinity whereas the shift at near covers only 5 cm.

cues signal increasing distance (Fried, 1973; Cormack, 1984). This relation is now known as Zuckerman's law.

Perceptual Instability After Biasing the VOR

We turn now to the perceptual sequela of the vestibulo-ocular response (VOR). As described in Chapter 3, the VOR underlies gaze stability and probably also what has been termed *visual position constancy* (Welch, 1978). By implication, when the gaze no longer is stable in space, perceptual instability should result.

Eliminating the VOR

One way to bias the VOR is to eliminate or reduce semicircular canal activity, typically by way of eighth-nerve lesion, or via the toxic-affects of strepto-mycin. In patients with vestibular dysfunction, maintaining fixation on nearby objects while walking is frequently accompanied by apparent movement of the environment. This comes about because of the failure of the VOR to compensate for the vertical head and trunk oscillations, typically between 0.6 and 8.2 Hz during walking and running (Grossman et al., 1988). Under normal conditions, both the upward head translation as well as nodding or rotational movements would induce a downward ocular rotation driven by the vertical VOR (Grossman et al., 1989) and perhaps by a transient otolithic reflex as well, thereby facilitating fixation on an earth-fixed object. In the absence of a compensatory mechanism, the eye would be carried along with the head and trunk in its vertical path while the pursuit and saccadic systems would have the burden of supporting the fixation reflex. Triggering the pursuit sys-tem probably is the dominant factor in accounting for the strong sense of object motion or oscillopsia reported by patients with vestibular dysfunction (e.g., J.C., 1952), but sheer retinal image movement may also contribute to this experience (Grossman & Leigh, 1990).

Initially this severe bias is quite debilitating, and even reading cannot be carried out without propping up the head. Fortunately, successful adaptation is possible based on the presence of visual input (J.C., 1952).

Atypical Stimulation

Another less drastic means of biasing the VOR is to stimulate it from without while the head is held stable. This can be accomplished by irrigating the external ear canals with either water or a thermally regulated stream of air. Typical water temperatures for the caloric test are 7 deg C below and above normative body temperature (i.e., 86 deg F and 111.2 deg F, respectively; Hood, 1984, p. 60). With the head pitched back so as to bring the horizontal canals into a vertical plane, then, presumably due to heat convection, the heat gradient within the endolymphatic fluid of the canal being stimulated provokes movement of the fluid and causes a cupula deflection along with nystagmus just as if the head had actually rotated to cause the same vestibular response (but only on the irrigated side). A typical perceptual response to caloric stimulation is indicated by reports of head and body rotation, and if the subject is fixating on a target, then the object also appears to move. Such target movement may occur during fixation while the nystagmus is overtly inhibited, and therefore no image movement is present. It is plausible therefore that the subjective motion observed under those circumstances is attributable to a pursuit signal that is used to cancel the slow phase of the thermally induced nystagmus. Therefore, it is as though one were pursuing a foveated target with no image movement on the retina. The underlying logic of this account is further developed in the next section. Generally, when tested in darkness, semicircular canal activity, regardless of the method of stimulation produces a sense of body rotation in the plane of the canal being stimulated (e.g., Mach, 1906/1959, p. 155; Israel, Sievering, & Koenig, 1995). Additional target motion percepts occur probably because of the associated nystagmus suppression by a balancing pursuit signal, when target fixation is permitted during canal stimulation. A similar explanation applies to the oculogyral illusion (Graybiel & Hupp, 1946; see also Chapter 5), an illusory target movement accompanying and following canal stimulation due to rotary acceleration of the body (Yessenow, 1972; Evanoff & Lackner, 1987).

Perceptual Effects of Adapting the Amplitude and Direction of the VOR

Studies of the adaptive nature of the VOR are not uncommon in contemporary vestibular research, some of which have been described in Chapter 3. Only a few, however, have sought to measure the perceptual consequences of these adaptations. For example, Gauthier and Robinson (1975) and Melville Jones, Berthoz, and Segal (1984) provide some evidence that an increase and decrease respectively in the VOR gain will be reflected directly in the perceived extent of body rotation in darkness. Thus, in principle, if the gain were to be reduced to zero [see Fig. 4.18(e)], then presumably no sense of body rotation would be reported as a result of radial acceleration. This interesting paradigm, comparing apparent self-rotation before and after VOR gain change, would seem to be a fruitful line of research. Are the changes in perceived self-rotation mediated by the effect of the vestibular apparatus on ocular muscle tonus and the ensuing eye movements, or on more central sites (Bloomberg et al., 2000)?

One frequently encountered outcome of VOR retraining is the instability of gaze that accompanies fixation of a target during head movement. Dubois

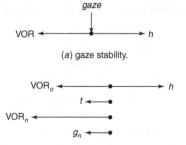

(a) gaze stability.

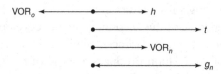

(b) increasing the velocity of the VOR (Gauthier & Robinson, 1975).

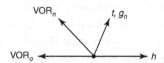

(c) reversing the direction of the VOR (Melvill Jones, 1977).

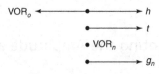

(d) shifting the direction of the VOR (Callan & Ebenholtz, 1982; Khater et al., 1990).

$$\begin{array}{c} \text{VOR}_o \longleftrightarrow h \\ \longrightarrow t \\ \bullet \ \text{VOR}_n \\ \longrightarrow g_n \end{array}$$

(e) reducing VOR velocity to zero.

Figure 4.18. *Gaze Instability after VOR Retraining.* (a) A head vector h, representing velocity and direction of head movement, when added to the VOR vector, representing eye-in-head movement, yields the trajectory of the eye in space, commonly termed *gaze*. When the VOR compensates completely for the head movement, the vectors cancel and gaze is perfectly stable. Furthermore, because the eye movements of the VOR are reflexive and not directly registered in consciousness, fixating a target while moving the head left to right typically yields a *percept* of a stable target in space. However, when the VOR and head movements occur in different directions and/or velocities, their vector sum is non-zero and gaze instability results. This is invariably the case after the VOR has been altered by adaptation. Each figure represents a different adaptation paradigm, following the vector model proposed by Dubois (1982). In each case, the subject pursues the target t during left to right head rotation. Because of the VOR, pursuit is not perfect, but engenders retinal slip signals that in turn are thought to cause changes in the parameters of the VOR (Robinson, 1976; Ebenholtz, 1984). In all instances, the new VOR (VOR$_n$) results from the vector sum of the old VOR (VOR$_o$) and the pursuit movement t, which also happens to be in the direction that minimizes the occurrence of slip signals. Furthermore, because the *(Continued on next page)*

112

(1982) devised a vector model that enables prediction of these directional instabilities. Following Dubois' proposal, analysis of four different VOR re-training paradigms (Ebenholtz, 1984b, 1986) are shown in Fig. 4.18. The recipe for retraining in all four cases is that a new VOR direction results from the vector sum of a pursuit or optokinetic movement made during a head rotation and the VOR that this head rotation normally triggers. That this yields a functional adaptation can be shown from the fact that the vector combination of the *new* VOR and the head movement causes the eye to move along a path that minimizes slip signals by keeping the eye fixated on the pursuit target. Figure 4.18 shows that this combination of VOR, which represents an eye-in-head direction, and a head movement (h) defines a new gaze direction in space (g_n) that is coincident with the pursuit-target path. The new gaze directions are represented in Fig. 4.19(b–e).

Once the new gaze vector is known, then the illusory movement of a spot fixated during a head movement can be inferred from the simple rule that if gaze is not stable in space, then in order to maintain fixation during a head movement, a pursuit signal must be issued in the antigaze direction. Presumably, the conscious representation of the pursuit signal is attributed to the fixated target, which is then perceived as moving.

Illusions of Motion and Extent Resulting From Pursuit, Saccades, and the Pursuit Suppression of the Optokinetic Reflex

Pursuit

When the eyes are in the pursuit mode and a target is being tracked, is there evidence that any perceived motion in this situation is attributable to the pursuit system? This is a deceptively simple question because there are three important technical criteria that must be met to reach a clear answer. First, there must be no movement of the tracked image on the retina for this by itself might produce a sense of movement. Second, the tracked image must not move relative to a fixed background since this might produce a sense of *relative* movement. Third, the eye must be in the slow pursuit mode, not performing saccades or merely fixating. Simply following a slowly moving target will not suffice because tracking is rarely perfect, so that retinal slip of the image on the retina frequently occurs, and also one typically finds

---

Figure 4.18. (*Continued*) new VOR and the head vector no longer cancel, there emerges a gaze instability represented by a new traveling gaze g_n.

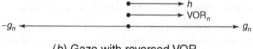

(a) Gaze with incresed VOR velocity.

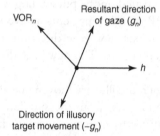

(b) Gaze with reversed VOR.

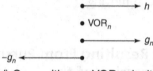

(c) Gaze with shifted direction of the VOR.

(d) Gaze with zero VOR velocity.

Figure 4.19. *Gaze Instability and Apparent Movement after VOR Retraining.* The direction of gaze in space is defined as the vector sum of head and VOR velocities. In the four cases illustrated in Fig. 4.18, the new or retrained VOR led to a non-zero sum, resulting in a new traveling gaze g_n. Presumably, only a target moving at the gaze velocity appears to be at rest. Here a truly stable target, fixated during a head movement, appears to move with vector $-g$ because of a pursuit signal necessary to cancel the gaze vector. Increased VOR velocity, as shown in (a), produced a gaze g_n that moved opposite the head rotation, and therefore, in order to retain fixation on a stationary target, a pursuit signal in the opposite direction $-g_n$ was required. Therefore, subjects reported movement of a target in the *same* direction as self-rotation (Gauthier & Robinson, 1975). With reversal of the VOR, as shown in (b), the new VOR and head vectors are equal and in the same direction; therefore, the new gaze is in the same direction as the head and at twice its velocity. The pursuit signal necessary to stabilize gaze is expected, therefore, to produce an illusory motion opposite the head direction. With a directional shift short of reversal, shown in (c), a rightward head rotation causes an oblique gaze shift g_n. Therefore, fixation on a stable target during head movement requires a pursuit signal $-g_n$, producing illusory motion down and to the left (Ebenholtz, 1986; Khater et al., 1990). Finally, if the VOR is reduced to zero, as shown in (d), so that the eyes make no movement relative to the head, *(Continued on next page)*

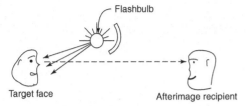

Figure 4.20. *Ghosts and Afterimages.* Any object that reflects light can be represented in an afterimage. All that is necessary is a sufficiently intense flash of light reflected from the target into the eyes of the recipient, as shown in the figure. When an afterimage is viewed in complete darkness, if the target object was colored, then initially so too is the afterimage, but quite desaturated. This positive phase is followed by a negative phase in which the brightness relationships in the original object are reversed. Positive and negative phases have been reported to alternate, and a "flight of colors" also has been reported in which the afterimage sequences through blue, red, orange, and yellow (Brown, 1965). Eventually the afterimage fades completely, but dark adaptation, blinking, and eye movements tend to prolong its life. The washed out colors, the nature of afterimages to be localized in the direction of gaze, and their intermittent quality qualify them as candidates for ghostly surrogates. One might imagine a serendipitous flash of lightening causing an afterimage in an uninformed recipient who then might claim a ghostly visitation. The method represented here produces reliable "ghosts" of friends and family.

saccades intermixed with the otherwise smooth pursuit movements. A partial solution is achieved with the use of an afterimage, produced by a brief flash of light as shown in Fig. 4.20. This is an excellent way to eliminate retinal slip because the afterimage simply represents ongoing retinal processes after the original stimulation has terminated. However, with no retinal slip it is difficult to engage the eye to shift into the pursuit mode. Nevertheless, observations by Helmholtz (1910/1962, vol. 3, p. 244) and others since clearly show that when the eyes move, the afterimage appears to move in the same directions. For example, Mack and Bachant (1969) recorded the apparent movement of a 4

Figure 4.19. (*Continued*) then it may be inferred that the gaze is carried to whatever direction the head assumes. This probably also describes the state of those without a functioning vestibular system, like J.C., described earlier. In order to maintain gaze on a stable object a balancing pursuit signal, $-g_n$ must be issued. This causes the stationary target to appear to move opposite the head direction but at an equal velocity. It is worthy of noting that suppression of nystagmus (VOR) plays a significant role in the training of figure skaters (Collins, 1966) and ballet dancers (Dix & Hood, 1969), and may prove to be a useful countermeasure for all forms of motion sickness (Ebenholtz, Cohen, & Linder, 1994). See the further discussion of motion sickness in Chapter 5.

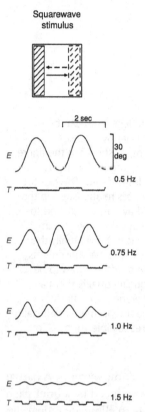

Squarewave
stimulus

Figure 4.21. *Apparent Motion of a Stabilized Image during Pursuit.* (Reprinted from Pola & Wyatt, 1989, The perception of target motion during smooth pursuit eye movements in the open-loop condition: Characteristics of retinal signals. *Vision Research, 29,* 471–483, Copyright 1989. Reprinted with permission from Elsevier Science.) The square, 3.75 deg side^{-1}, was stabilized symmetrically at the fovea while the vertical bars, 1 deg wide, moved *instantaneously* from the left to the right side of the fovea over a range of frequencies. Thus, neither the image of the stabilized square nor of the bars produced any slip signals on the retina. Under instructions to "try to look at the jumping bar" (p. 473) and also to "observe the amplitude of the motion of the square" (p. 473), the eyes moved in sinusoidal fashion typical of slow pursuit movements. The sinusoidal eye movement recordings *E* are shown below the target, whereas the square wave record shows the signal driving the bar target *T*. Because both the square and bars were stabilized, they stimulated the eye at fixed retinal locations regardless of eye movements; therefore, any apparent movement of the outer square was due to the registered movement of the eye itself. In order to measure the apparent length of the movement path of the outer square, subjects were shown a nonstabilized truly moving square whose amplitude was adjusted to match that of the previously seen open-loop target. Overall, the *apparent* amplitude (*Continued on next page*)

deg afterimage by tracing its path with the unseen right hand of the observer, which served as pointer. The authors reported reasonably well-correlated eye and pointer records of horizontal movements, which included both saccades and slow drifts.

Similar reports of apparent movement of a small afterimage of about 20 min of arc were provided by Pelz and Hayhoe (1995), under instructions to maintain the direction of gaze in the dark. Eye movement recording showed both slow drift and saccades along with a perception by the observers that both the afterimage and their eyes had moved. Pelz and Hayhoe also made corresponding observations with a large 45 deg extended afterimage with the interesting result that although larger spontaneous eye movements occurred up to about 13 deg from the original position, the afterimage scene appeared stationary. These results imply that large afterimages play a role in biasing or suppressing the eye position signals that otherwise would convey eye and afterimage movement. It should be borne in mind that although this conclusion applies to small saccades and slow drifting movements, it does not apply necessarily to pure pursuit movements as such.

Before describing research that actually seems to have studied pursuit movements without slip signals, it is worth noting that several researchers have measured the perceived velocity of a target during tracking and also the apparent length of the movement path (Mack & Herman, 1972; Coren, Bradley, Hoenig, & Girgus, 1975; Festinger, Sedgwick & Holtzman, 1976; Miller, 1980). Overall, these contemporary studies show that the perceived eye movement path during pursuit typically is shorter than the same path traversed by saccades, and that the apparent velocity of pursuit is low and nonveridical. It has long been known that a target seen during tracking appears to move slower than it does while the eye remains stationary and fixed (Fleischl, 1882). This phenomenon, the Aubert–Fleischl effect, may be related (as suggested by Mack and Herman, 1972) to the underestimation of the movement path of a tracked target.

One of the most elegant studies of perceived target movement during pursuit was that of Pola and Wyatt (1989), who achieved the aim of examining pursuit movement without retinal slip through the use of open-loop tracking. The square pursuit target, shown in Fig. 4.21, was optically stabilized on the

←——————————————————————————

Figure 4.21. (*Continued*) of target movement was less than the actual eye movement path and diminished steadily with increasing target frequency. For one subject, underestimation ranged from 80 percent to 20 percent of eye movement path. Thus, although pursuit movements do convey the length of the eyemovement path, they do so with considerable underestimation.

retina as in the afterimage studies. However, because smooth pursuit eye movements were induced by attempts to look at the flashing bars, as shown in the figure, Pola and Wyatt reasoned that perceived movement of the square must be attributed to the conscious registration of ocular movement. Presumably, this perception is mediated by either or a combination of extraretinal neural signals used to drive the eyes (efference) or signals from the eye muscles themselves (afference). Is the perceived movement mediated by the pursuit system veridical? Measurement of the perceived movement path showed unambiguous underestimation of the actual eye-movement amplitude in reference to a closed loop measurement procedure that allowed retinal slip signals to occur. There is thus excellent concurrence among studies showing that although pursuit movements convey a sense of eye-movement path, they generally are nonveridical (Mack & Herman, 1972; Coren et al., 1975; Honda, 1990; Mateeff, Hohnsbein, & Ehrenstein, 1990). Furthermore, the longer one pursues a target, the more likely one is to underestimate its path length (Miller, 1980). This is probably due to a robotizing or adaptive effect of prolonged practice with pursuit movements (Wallach, Schuman, & O'Leary, 1981). The more automatic the movement becomes, the less able one is to extract conscious correlates of the movement path.

It is not clear why the pursuit system but not the saccadic system yields underestimation of path length. In addition, because apparent length requires some input concerning target distance, it would be interesting to know whether distance information is suppressed during pursuit eye movements but available while saccading. Is there some degree of size constancy for apparent length mediated by the pursuit system?

Saccades

The studies described previously suggest that saccades, interspersed with pursuit movements, tend to mediate veridical perceptions of path length, but as in the case of the pursuit system, detailed studies of apparent path length as a function both of target distance and saccade amplitude are lacking. Therefore, it remains to be shown whether there exists a size constancy for saccadic path length. It is known, however, that an afterimage appears to displace during a saccade, in the direction of gaze. This was described in the third century BCE by Aristotle, who, while observing the characteristics of positive after images, noted that "if, after having looked at the sun or some other brilliant object, we close the eyes, then, if we watch carefully, it appears in a straight line with the direction of vision (whatever this may be)" (Ackrill, 1987, p. 214). It may

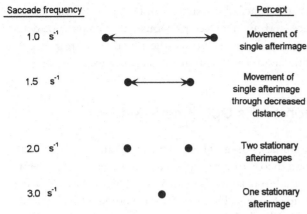

Saccade frequency		Percept

1.0 s^{-1} — Movement of single afterimage

1.5 s^{-1} — Movement of single afterimage through decreased distance

2.0 s^{-1} — Two stationary afterimages

3.0 s^{-1} — One stationary afterimage

Figure 4.22. *Perceived Movement of an Afterimage during Saccades.* (Modified from, Grüsser, Krizic, & Weiss, 1987, Afterimage movement during saccades in the dark. *Vision Research, 27,* 215–226, Copyright 1987. Reprinted with permission from Elsevier Science.) An afterimage from a high intensity source, lasting about 2 to 5 min, was tracked in darkness by moving a handle in a circular track to the left or right of the subject's median plane. Saccade rates were controlled by two 500 Hz square wave signals fed into two loudspeakers placed at eye level about 140 cm from the subject, and at 19.5 deg left and right of straight ahead. Subjects were asked to look toward each sound as quickly as possible while eye movements and handle position were monitored. As the saccadic frequency increased, the actual amplitude stayed relatively constant, whereas the afterimage, as tracked by the handle position, underwent drastic changes. Beyond about three saccades s^{-1} only a single stationary afterimage was seen, approximately in the subjects midline. These results graphically indicate that even though reliable saccades are carried out in the frequency range examined, the system fails to extract information about changes in eye position except at relatively slow movement rates.

not be surprising that an image impressed on the fovea would be seen in the direction of gaze, but in order for this to come about, the position of the eye must be available to centers of consciousness. It follows that if the eye changes position too rapidly for the brain to process the changed locations, the sensed position of the afterimage also deteriorates. Figure 4.22 shows the results of Grüsser, Krizic, and Weiss, (1987), in which the voluntary saccadic frequency was varied while subjects viewed and manually tracked the changing location of the afterimage in space. They found that beyond about two saccades per second the sense of movement was lost. Subjects then experienced either two afterimages simultaneously or at still higher frequencies, a single afterimage directly in front of them. Even at low saccade frequencies in which a single afterimage was seen to move with the eye, the length of the movement path diminished as the frequency changed from one to two saccades per second.

We may conclude that the saccadic system is not necessarily the gateway to veridical performance, and, as suggested previously, it remains to be determined whether, and if so, how, distance and saccadic information are integrated to evaluate saccadic path length for targets at various distances.

Pursuit Suppression of the Optokinetic Reflex

In a number of creative studies, Post and Leibowitz (1985) and colleagues have developed evidence for a theory of motion perception based on the premise that a pursuit signal is used to balance some other oculomotor drive such as that from an optokinetic stimulus. The approach is very similar to that developed independently by Ebenholtz (1986) to account for illusions resulting from adaptation in various oculomotor systems. In these instances, in order to maintain fixation, it frequently was the case that a voluntary oculomotor signal had to be issued so as to balance an opposing reflexive system that, if unopposed, might pull the eye off target, thereby preventing fixation. Examples are a convergence signal to balance against a vergence hysteresis effect (Ebenholtz & Fisher, 1982), a pursuit signal against a VOR (Ebenholtz, 1986; Post & Lott, 1992), or a pursuit signal against an OKR (Post, Shupert, & Leibowitz, 1984; Heckmann & Post, 1988). In all cases, the voluntary signal predominates in conveying a spatial attribute such as distance or motion, whereas the reflexive component remains suppressed and provides no conscious correlates. A wonderful example of this approach, proposed by Post and Leibowitz (1985), is the explanation of the *traveling-moon illusion*. In a typical case, the illusion would occur when a passenger in a car observes the moon by viewing out of his or her side window. On a straight road, the forward path of the vehicle produces only a minute change of the direction of gaze, owing to the 238,857 miles of distance to the moon and the short path of the vehicle on the Earth's surface. Hence, one might expect the resulting stationary gaze direction to give rise to the sense of a fixed nonmoving moon. Yet, the typical experience is that the moon travels *with* the car, in good correspondence with its velocity. Consistent with Post and Leibowitz (1985), and Ebenholtz (1986), the presence of tree-tops or roof-top contours may be presumed to serve as an optokinetic stimulus to rotate the eyes opposite the direction of travel. Therefore, a countervailing pursuit signal would be required in order to maintain fixation on the moon. Because the illusion may be obliterated by eliminating sight of the landscape, the role of the OKR is confirmed. Furthermore, because a pursuit signal is necessary to balance the action of the OKR, the eye itself does not move, but the sense of movement is provided by the pursuit signal itself. The same oculomotor mechanisms underlie the classic phenomenon of *induced movement* (Duncker, 1929), whereby a fixated, stationary spot appears to move opposite the direction of a truly moving surrounding rectangle (Post, Shupert & Leibowitz, 1984).

Illusory Percepts After Vibrotactile Stimulation of Extraocular Muscle

When the spindles of skeletal muscle are stimulated by tendon vibration (Goodwin, McCloskey, & Mathews, 1972b; Craske, 1977), an induced contraction occurs, as shown in Fig. 3.14(b). If the contracting limb is constrained from moving, then increasing spindle activity signals as if the limb were extending and stretching the stimulated muscle. Assuming the biceps muscle was initially stimulated, the subject experiences the arm rotating around the elbow joint hyperextended in the "wrong direction." A similar experience occurs with stimulation of a wrist flexor muscle (Craske, 1977). These paradoxical limb-position illusions demonstrate that the perceptual interpretation of proprioception is not simply determined by past experience, because these limb positions never occurred in the life experience of the subjects. They do, however, lend themselves, as Craske suggests, to the possibility of extrapolation from past experience, as though solving an equation by substituting values for the unknowns never before tried.

Given the effects of vibratory stimuli on perceived limb position, the reasonable question emerges about the effects of vibratory stimulation on extraocular muscle, where stretch-detecting muscle spindles also are known to exist (Cooper & Daniel, 1949; Whitteridge, 1959). In an ingenious series of studies, Roll and colleagues (1987; 1991) explored this issue by developing and applying a minielectromagnetic vibrator periorbitally over the surrounding skin at sites corresponding respectively to the lateral and medial recti and the inferior and superior recti. An example of vibrator placement to stimulate the right inferior rectus is shown in Fig. 4.23. In this case, when the subject fixated a small LED (Roll, Velay, & Roll, 1991), an illusion of upward target movement was reported. In other studies (Roll & Roll, 1987) the induced movement in the target also was shown consistently to be in the direction opposite the vibrated muscle. Roll and Roll report that as a general rule "the direction of . . . illusory . . . displacements was always that which corresponded to the stretch of the vibrated extra-ocular muscle" (p. 65). Although not stated as such by the authors, an underlying mechanism may be adduced that is consistent with

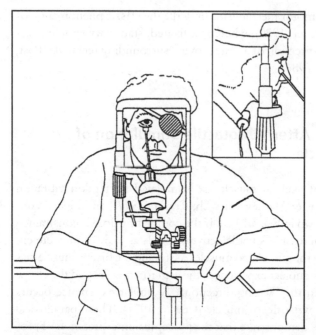

Figure 4.23. *Periorbital Vibration of Extraocular Muscles.* (From Roll, Velay, & Roll, 1991, Eye and neck proprioceptive messages contribute to the spatial coding of retinal input in visually oriented activities. *Experimental Brain Research, 85,* 423–431. Copyright 1991 by Springer-Verlag. Reprinted with permission from Springer-Verlag GmbH & Co. KG.) Vibration with a 3 ms rectangular pulse at an amplitude of 0.1 to 0.2 mm was applied for about 3 sec. Subjects viewed an LED during stimulation, and when it was over pointed in complete darkness to the place in space it appeared to occupy just before it disappeared. The pointing finger contacted an electronic panel that recorded the spatial position of the finger. In this manner apparent displacements caused by illusory movement of the LED were tracked manually. The figure shows the vibrator positioned to stimulate the inferior rectus of the right eye. Maximal apparent displacement occurred at a vibration frequency of about 80 Hz. At this frequency the target appeared to shift upward about 4.5 deg from the position measured without vibratory stimulation. In general, the fixated target appeared to move in a direction opposite the normal field of action of the stimulated muscle.

the rule offered by Roll and Roll. Assume first, that vibration of extraocular muscle causes a reflexive contraction in the vibrated muscle. This would be expected to cause the eye to rotate toward the direction of the stimulated muscle, were it not for the sustained fixation on the LED. When a fixation target is present, perhaps owing to a compensating pursuit signal, the eye does not actually move, and therefore the continued vibratory stimulation of the spindles is not relieved by a muscle contraction. Because the muscle-spindle induced contraction is counterbalanced or suppressed by the pursuit signal, the gaze is registered as if it were moving opposite the direction of the vibrated muscle. Likewise, the fixated target appears to move in the same direction. The validity of the postulated pursuit signal is supported by the observation that with the eyes closed, vibration produced no sensation of eye movement (Roll & Roll, 1987). Therefore it would appear that the pursuit signal is the carrier of the movement percept when a target is fixated.

Roll and colleagues also found that extraocular muscle vibration induced head, trunk, and whole-body movement illusions. Some of the implications of these findings are discussed in Chapter 5.

To many students of perception, especially those of us trained in the tradition of the Gestalt psychologists, it is startling to first consider the large number of oculomotor systems that contribute to perception as well as the large array of perceptual attributes mediated by them. Many questions of a general nature arise out of the view that oculomotor systems, directly and indirectly, determine certain aspects of perception. Chapter 5 takes up some of these general theoretical implications and unanswered questions. In addition, a theoretical basis is offered for the possible role of oculomotor systems in some very important phenomena, such as motion sickness and vection.

5

Theoretical Issues and Underlying Mechanisms

Introduction

Just how perceptions get to be triggered along with certain motor acts will eventually be answered when the underlying neural networks are better understood. For now, it is clear that although certain oculomotor systems such as convergence and pursuit are associated with conscious states, others such as the VOR and OKN are not. Furthermore, the circumstances under which these systems become active frequently are indirect and not at all apparent. The compensation theories discussed in this chapter provide good examples of how one oculomotor system comes to be turned off or suppressed while another is substituted with frequently surprising perceptual results. Likewise, compensating for the effects of oculomotor adaptations although not requiring substitution of one system for another requires voluntary internal changes in innervation level or directional control. In this way, they also provide the occasion for surprising perceptual outcomes.

In addition to compensating activity, other modulating factors play a role in determining the magnitude of perceptual effects associated with oculomotor systems. Some of these discussed here are ambient light levels, voluntary and reflexive oculomotor activity, and transient vs. steady-state eye movements. Additional topics include the role of eye movements in motion sickness and the remarkable phenomenon of induced self-movement (vection).

Compensation Theories: Role of Reflexive and Voluntary Eye Movements in Normative Perception and Illusions

Early in the development of the modern period of vision sciences, eye movement effort was recognized as a significant factor in controlling egocentric location and apparent movement. For example, Helmholtz (1910/1962, Vol. 3, p. 245) reported of a patient with a paralyzed right lateral rectus, that a target appeared to move to the right when the patient attempted, unsuccessfully, to look to the right by rotating the fixating eye. The eye itself did not move because of the paralysis, so the changed perceptual direction could properly be attributed to the effort, or to efference, issued to the lateral extraocular muscles. Similar observations, now termed *past pointing*, were known to Mach (1906/1959, p. 128), who devised a heroic experiment he performed on himself. He hypothesized that individuals with right-lateral muscle paralysis reached too far to the right when attempting to grasp an object because the extra effort required to move the eyes caused the target to appear to undergo a rightward displacement. He tested this experimentally by inhibiting movement of his eyes while attempting a voluntary rightward movement. His description follows:

> Let the eyes be turned as far as possible towards the left and two large lumps of moderately hard putty firmly pressed against the right side of each eyeball. If, now, we attempt to glance quickly to the right, we shall succeed only very imperfectly, owing to the imperfectly spherical form of the eyes, and the objects will suffer a strong displacement to the right. Thus the mere will to look to the right imparts to the images at certain points of the retina a larger "rightward value." (p. 128)

Mach never informed us as to whether his procedure produced bloodshot eyes or an otherwise damaged conjunctiva, but the attempt to move from clinical observations of cases of ocular paralysis to controlled instances of simulated paralysis clearly was in the spirit of the development of modern science and of psychology as an empirical science.

Apart from the philosophy-of-science implications, Mach's experiment and the clinical observations on individuals with paralyzed eye muscles both support the same conclusion; namely, that the illusions described arise directly out of the voluntary issuance of ocular innervation in an effort to compensate for ocular muscle inadequacy. In an unresponsive eye due to partial paralysis, atypical levels of innervation required to put the eye on target itself produces illusions of movement or displacement.

Position Constancy

If a target image changes its retinal location with each voluntary shift in gaze, why then do we not experience a constantly shifting visual world with each saccade? Mach framed this question, known today as the problem of position constancy, and also proposed an answer. He noted that spatial location could not be determined unambiguously by retinal stimulation alone, and from his ocular putty experiment and the past-pointing observations he concluded that the eye muscle innervation pattern together with the retinal image location jointly conveyed the spatial location of objects. We have already seen in Chapter 4 that if an image is fixed on the retina, then a saccade causes it to be seen wherever the observer fixates. Accordingly, the underlying logic for an explanation of position constancy would seem to run as follows: If images of the surroundings did not move or change locations on the retina as the eye moved, then they would actually appear to shift, with each voluntary eyemovement. Therefore, retinal image movement or displacement *in* the direction of a voluntary eyemovement is necessary to balance against the change in apparent spatial location that would otherwise occur if images were fixed in place on the retina. In short, it is the systematic displacement of images on the retina during a voluntary eyemovement that assures position constancy in normative perception.

A more general model of the position constancy process has been developed by von Holst and Mittelstaedt (1950). It too is based fundamentally, on the premise of a neurologic trade-off between image-movement afference and eye movement efference. In this approach, each efferent command is presumed to generate a neural copy, which in turn is to be compared with the afferent neural stream, termed *reafference*, produced as a direct result of the moving effector. For example, if the effector is an eye muscle then the reafference might arise from retinal image movement. If the comparison between efference and reafference leaves no residual, then an equilibrium state may be presumed and perceptual stability will be experienced. In contrast, if the efference is left uncancelled by reafference, as in the case of an unmoving paralyzed eye, then the perception follows the efferent command, and the same result occurs when in a normal eye the eye movement occurs with a retinal afterimage in place. When afference occurs from some external source that, therefore, is not tightly correlated with some efferent command, it is termed *exafference*. Then, as in the previous cases, any residual from the afference–efference comparison results in illusory percepts.

Sources of Sentience

Since the time of Helmholtz and Mach, numerous studies and conjectures have been developed on the matter of ocular muscle proprioception and the general question of the contribution of efference to perception (e.g., Festinger, Ono, Burnham, & Bamber, 1967; Shebilske, 1987). Although new research has helped to define the issue, the basic contributions of Mach have remained at the core of modern treatments. The central concepts are (1) that voluntary, but not reflexive innervation leads to changes in space perception; and (2) that in many instances, the subtle substitution of voluntary for reflexive innervation results in illusions of spatial location and/or movement. These substitutions generally are compensatory for the potential loss of fixation that reflexive innervation such as the vestibulo-ocular response (VOR) and optokinetic reflex (OKR) might otherwise produce.

On the matter of perceptual effects of compensating for the VOR, Mach (1906/1959) described the hypothetical state of an observer in a rotating room.

> If an observer be shut up in a closed receptacle, and the receptacle be set in rotation, toward the right, it will appear to the observer as optically in rotation, although every ground of inference for relative rotation is wanting. If his eyes perform involuntary [slow phase] compensatory movements to the left, the images on the retina will be displaced, with the result that he has the sensation of [object] movement toward the right. If, however, he fixes his eyes upon the receptacle, he must voluntarily compensate the involuntary movements, and thus again he is conscious of movements towards the right. (p. 138)

Presumably, the experienced movement is based on a rightward directed pursuit signal issued to compensate for the leftward-acting VOR.

Mach's theorizing about the perceptual effects of compensating for OKR also by means of an oppositely directed pursuit eyemovement signal is represented in the following observation:

> Objects in motion exert, as is well known, a peculiar motor stimulus upon the eye, and draw our attention and our gaze after them. If the eye really follows them, we must assume ... that the objects appear to move. But if the eye is kept for some time at rest in spite of the moving objects, the constant motor stimulus proceeding from the latter must be compensated by an equally constant

stream of innervation flowing to the motor apparatus of the eye, exactly as if the motionless point on which the eyes rest were moving uniformly in the opposite direction, and we were following it with our eyes. (p. 146)

It seems clear from Mach's observations that ocular sentience, the state of awareness of direction and movement of the eyes, derives from the application of voluntary eyemovement systems that substitute for reflexive ones. The latter, such as optokinetic nystagmus (OKN) and VOR, presumably have no conscious correlates, but just why reflexively driven eye movements should not provide a conscious record is a matter of speculation. For example, Ebenholtz (1986) attempted to explain why voluntary eye movements, made after adaptation of the direction of VOR and after adaptation of convergence, led to illusory perceptions. He proposed a mechanism similar to that discussed by Matin (1976) in the context of visual-direction illusions associated with the "paralyzed-eye paradigm" described in a later section. According to this proposal, the degree of sentience was related to the amount of γ-efferent stimulation of the extraocular intrafusal muscle fibers. These small fibers lie within the larger muscle fibers and are supplied with specialized muscle spindles or sensors that output a signal when stretched. It is known that human extraocular muscle is well supplied with muscle spindles (Cooper, Daniel, & Whitteridge, 1955), and therefore it is theoretically possible that the output of these spindles signals eye position or trajectory. Figure 5.00 provides a stylized representation of how the γ-efferent system modulates spindle output (Whitteridge, 1959), even when the larger surrounding muscle itself has not been stretched. The critical feature of this approach, however, is the presumed link between volition and heightened levels of γ-efferent stimulation (Merton, 1972) on the one hand and the absence or diminished level of γ-efferent stimulation during reflexive or passive activity on the other hand. Because the γ-efferent system boosts the output of the muscle spindles, doing so only during voluntary control would ensure greater sentience for eye position after voluntary innervation.

Typically, the initial command to fixate or keep pursuing a target is voluntary whereas the rest of the eye movement proceeds without awareness. Because all voluntary eye movements are combinations of voluntary and reflexive components, Ebenholtz (1986) proposed that it is only the steady-state portion of the eye movement that is associated with voluntary activity and therefore is manifest in conscious centers. In the case of a saccade, for example, one is not conscious of the transient pulse phase, whereas the steady-state step phase of the response yields the sense of gaze direction. See Chapter 6 for a caveat on this principle.

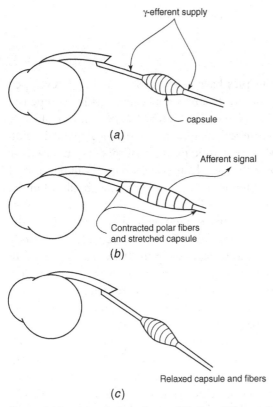

γ-efferent supply

capsule

(a)

Afferent signal

Contracted polar fibers
and stretched capsule

(b)

Relaxed capsule and fibers

(c)

Figure 5.00. *Muscle Spindles and Their γ-Efferent Stimulation in Extraocular Muscle.* The γ-efferent supply to the intrafusal fibers along with the α supply to the larger extrafusal muscle fibers are thought to be implicated in voluntary control (Granit, 1970). The γ-efferent mechanism in Fig. 5.00(a) is one of two types that operate by stimulating contraction in the intrafusal fibers on both sides of the capsule containing sensory nerve endings. Contraction of the spindle fibers, shown in Fig. 5.00(b), causes the capsule to stretch and a sensory signal proportional to the degree of stretch to be sent centrally (Steinbach, 1987). As shown in Fig. 5.00(c), subsequent contraction by α-innervation of the larger muscles (partially represented) that surround the intrafusal spindles would actually rotate the eye upwards thereby relieving the tension across the intrafusal spindles. In general, voluntary eye movements may occur by way of the γ-efferent system, first by causing the spindle capsule to stretch and then by way of reflex innervation of the larger muscle group in order to remove the tension across the capsule (Merton, 1972).

It is only a theoretical possibility that α-γ coactivation actually leads to awareness of eye position in the orbit. By inference, passive eye movements occurring without α-γ coactivation would not lead to sentient eye positioning, nor would reflexive eye movements, occurring via direct α innervation, provide a sense of eye movement or position. Empirical studies of coactivation of human extraocular muscle are needed to evaluate these theoretical possibilities. Does sentience arise out of the monitored muscle spindle afference, or out of monitored efference?

Oculogyral Illusion

In modern times, Mach's principles have been rediscovered and developed (e.g., Bruell & Albee, 1955), and creative new applications of his compensation theory have been described (Whiteside, Graybiel, & Niven, 1965; Post & Leibowitz, 1985). Recent theoretical treatments of the oculogyral illusion (OGI; Graybiel & Hupp, 1946) are a good example of modern applications. The OGI occurs when an observer is accelerated in the dark about the vertical body axis while fixating a small luminous target placed straight ahead in the observer's median plane. Both the target light and the observer rotate and accelerate together, so there is no objective motion or displacement between the two. Yet the target light is seen to move and displace in the direction of body rotation. For moderate rotary acceleration rates up to 30 deg/sec^2 (Evanoff & Lackner, 1987), fixation on a target effectively suppresses nystagmus. Accordingly, Whiteside et al. (1965) proposed that the apparent displacement was due to the fixation effort needed to compensate for this nystagmus that otherwise pulls the fovea off the target image. In a similar analysis, Post and Leibowitz (1985) proposed that fixation was maintained by the pursuit system which was voluntarily activated so as to oppose the reflexive nystagmus or VOR. Both analyses may be correct because fixation typically occurs when the pursuit system is triggered. It remains to be noted that the compensation-theory explanation of OGI has been validated by the ingenious study of Evanoff and Lackner (1987). By varying the angle of the fixation target in relation to the observer's median plane, they modulated the slow phase amplitude and velocity of an induced rotary (horizontal) nystagmus. They inferred that the compensatory pursuit signal necessary for continued fixation would be modulated in like fashion, and found that the illusion magnitude indeed was altered predictably.

Light vs. Dark Environments

There is a somewhat controversial literature (Turvey & Solomon, 1984; Shebilske, 1987) stemming from the theory that perception may be direct, driven by information in the stimulus pattern (Gibson, 1979), but unmediated by underlying mechanisms. The question whether oculomotor-based changes in perception occur in normally lit environments is theoretically interesting because, in such contexts, visual-pattern information frequently exists for specifying target location, and so oculomotor cues may be outweighed or simply inhibited. To examine this question, a number of investigators have undertaken the measurement of oculomotor-based illusions in both darkness and well-illuminated, full-cue conditions. Do oculomotor systems contribute to perception in structured environments?

The Paralyzed-Eye Paradigm

When eye muscles are paralyzed, innervation that would normally contract the extraocular muscles fails or does so only partially. As a consequence, the eye either does not respond at all to voluntary commands or gaze falls short of the intended location and requires a corrective saccade. Subjectively, the latter condition is experienced as a jumping and displacement of the visual scene in the direction of the eye movement, and observers typically point beyond the actual target location (Stevens et al., 1976). These results are consistent with the classic observations of Helmholtz and Mach on clinical cases of ocular paresis and also with the observations of subjects in modern heroic studies of induced paralysis produced by the administration of curare (e.g., Stevens et al., 1976; Matin et al., 1982). Overall, studies of induced partial paralysis may be interpreted to imply that the illusory perceptions are due to the exaggerated efferent signal necessary to contract the partially paralyzed eye muscles to bring the target image onto the fovea.

Although the apparent displacements reported by Stevens et al. (1976) were experienced in an illuminated environment, consisting of a fixation point on a 7 × 10 ft screen at about 7 ft, no quantitative measures were made nor were control observations made in darkness. These potential deficiencies were remedied by Matin et al. (1982). In that study, five subjects were administered injections of D-tubocurarine (commonly referred to as *curare*) to the point where expiratory capacity was reduced to 85 percent of normal, although machine-assisted breathing was not required. Matin et al. reported that "speech was barely audible, the arm could barely be raised from the lap" and "ptosis [lid droop] and diplopia were present throughout"(p. 201); therefore, subjects viewed monocularly only. The partially paralyzed observers made several types of systematic observations in both darkness and a lighted environment. In the latter case, subjects viewed a target in a rectangular room, about 25 × 35 ft, from a slightly pitched-back position on a surgical chair, with their feet visible in front of them. The room contained several tables supporting laboratory equipment and the walls were lined with glass-doored laboratory cabinets, (L. Matin, personal communication, 1997). The room thus contained a good deal of visible structure. In one task a 24 × 24 min of arc transilluminated *E* was fixated while the subject signaled the experimenter by making grunts to adjust a second target so that it appeared at eye level (i.e., at a vertical location in space corresponding to the horizon where gaze direction would be horizontal). Measurements of apparent eye level taken for several vertical locations of the fixation target showed that in darkness, the lower the fixation target, the higher the adjustable target had to be placed in order to appear at eye level. Likewise, high fixation targets caused low eye-level settings. Analogous results occurred when curarized subjects set a target to appear in the median plane while fixating a second target placed at various positions to the left and right of straight ahead. The more rightward the fixation target, the more leftward the adjustable target was positioned in order to appear in the observer's median plane, and left fixation targets produced rightward-displaced median plane targets. Both the vertical- and horizontal-adjustment tasks produced strong illusions, but only in darkness. Matin, Stevens, and Picoult (1983) noted that, "in normal illumination, visual localization is extremely accurate" (p. 255).

The results in the dark environment are consistent with the perception of subjects whose weakened eye muscles caused them to issue abnormally high levels of innervation in order to maintain target fixation. Because high levels of innervation would signal large target eccentricities, then the adjustable targets would have to be displaced in the opposite direction to a correspondingly large extent.

Other observations also support the premise that abnormally high innervation levels are required in the curarized state in order to maintain fixation. For example, Matin et al. (1983) report that when viewing a target truly set at eye level and with the head tipped back, thus requiring a downward gaze deviation, then when lights were turned out the target "appeared to move slowly in the direction of the (invisible) floor" (Matin et al., 1983, p. 245) and to appear immediately at eye level when the room was reilluminated. Similarly, when the head was tipped forward, requiring an upward elevation of gaze to maintain fixation, darkening the room caused the target, "to rise to a position near the (invisible) ceiling" (p. 247). But why were the effects of partial paralysis not apparent in full illumination?

Several answers are available. First, Matin et al. (1983) showed that when a sound was matched to a horizontally displaced visual target, the displacement illusion was present (i.e., the sound chosen was from a loudspeaker displaced to the side of the fixation target toward which the eyes were turned), and this was true equally for both lighted and dark environments. The effects of partial paralysis thus indeed were present in full illumination, provided they were measured by the sound localization technique. Second, median plane and eye-level adjustments made with visual targets did not show

illusory displacements in the illuminated room. The reasons for these equiv-
ocal results are not clear. However, Bridgeman (1996) proposed that the
darkening of the room may have helped to remove a source of reflexive innervation, stemming from the visual pattern of the room, which may have
helped to maintain fixation without the added voluntary innervation required
of the weakened eye muscles in darkness. In other words, the room may
have provided a significant *optostatic* stimulus (Crone, 1975), a static visual
pattern influencing the direction of gaze, making it easy to maintain fixation.
This would be a viable explanation if the visible structure of the room was
altered (reduced) to make it less of an optostatic stimulus during the auditory measurement phase. There is no evidence, however, that this was the
case.

It seems clear from the studies of Stevens et al. (1976) and Matin et al. (1982)
that illusory displacements are experienced by partially paralyzed observers,
even in illuminated environments. It remains unclear, however, why the illusion
could not be detected in the nonauditory part of the Matin et al. (1982) study,
but as Matin et al. (1983) noted (p. 258), the readily available sight of self
could have played some role to this end by providing a visible reference for
alignment of the target light with some aspect of the body.

In addition to observing the effects of the systemic injection of curare, a
number of courageous investigators also examined perceptual effects following anesthetization of five of the six extraocular muscles, sparing the superior
oblique. Stevens et al. (1976) are the most recent investigators to have undergone retrobulbar block, by infusing procaine behind the eyeball. Under
total block, past pointing was done uniformly, but subjects reported that the
apparent target displacement thought to underlie the past pointing, "was not
necessarily visual in nature, but simply the feeling that if you wanted to touch
a given object you would have to reach to the right" (p. 97). A similar report was made after total ocular muscle paralysis was achieved via curare.
In contrast, under partial curare-induced paralysis or before the onset of full
retrobulbar block, perceptual displacements of the scene in the direction of
the eye movement were experienced (Stevens et al., 1976). Thus, the perceptual
effects of eye-muscle paralysis may vary, depending on the extent of the paralysis. It may be relevant here to consider the experience of a normal observer
whose eye is rotated to its extreme position. Any scene viewed, with no matter
how great an effort, does not appear to have moved or to be displaced. It is
possible that this condition corresponds to the situation of complete ocular
paralysis, and conversely that some degree of muscle response or muscle-
spindle activation is required in order to experience illusory displacement and
motion.

The Eye-Press Paradigm

This paradigm (see Chapter 4) was first described by Descartes (1664/1972) and carefully investigated in recent times by Rine and Skavenski (1997) and extensively by Bridgeman and colleagues (e.g., Stark & Bridgeman, 1983; Ilg et al., 1989). With one eye covered, finger pressure on the side of the open eye produces a lateral shift (Stark & Bridgeman, 1983) of the eyeball. However, it is possible to maintain fixation on a target even though the eye is being displaced in its orbit. This is accomplished by having the subject make a slight rotation of the eye opposite the finger pressure toward the target, and is represented in Fig. 5.01. In this maneuver the compensatory eye movement and the underlying innervation to the extraocular muscles must be such as to make up for both the ocular displacement caused by the finger press as well as the increased friction or pressure imparted by the finger pressing on the lid. According to Hering's law of equal innervation (see Chapter 3) the same level of efferent signal is sent to both eyes, and, in fact, in the covered eye that was free to move Stark and Bridgeman (1983) found large compensatory deviations opposite the direction of the eye press.

Evidence for the role of this compensatory innervation of extraocular muscle in perception was readily found when subjects pointed at a luminous target with their unseen hand or matched the position of an auditory signal with a visual target. This was true in both darkness and in the presence of a fully lit visual field. Only in the case of visual measures made in a structured visual field did the compensated eye press fail to induce an apparent shift in position of the target. For example, in the dark, a target adjusted to appear in front of the pressed right eye (parallel to the subject's median plane) actually was positioned to the left, opposite the rightward gaze conveyed by the efferent signal that compensated for the pressed-eye displacement. This correction amounted to about 2 deg, whereas in the light virtually no error occurred.

It seems clear that, even in an illuminated, structured visual field, compensatory ocular innervation issued to overcome either partial paralysis or finger pressure on the eye governs the perception of visual direction. A major caveat however is that not all measures of perception reflect this fact. Both Matin et al. (1982) and Stark and Bridgeman (1983) are consistent in showing that a structured visual environment suppressed or outweighed the effects of compensatory eye muscle innervation when visual targets were adjusted to match some criterion such as "eye level" or "straight ahead." Some support for the optostatic interpretation of these null effects in structured, lighted environments has been provided by Bridgeman and Graziano (1989), but it is not

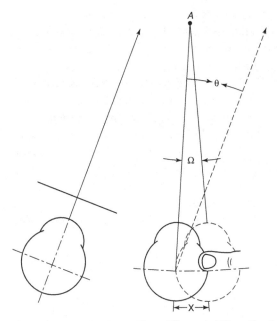

Figure 5.01. *Eye Press Alters Efference When Fixation Is Maintained.* (Modified from Stark & Bridgeman, 1983, Role of corollary discharge in space constancy. *Perception & Psychophysics, 34,* 371–380. Reprinted with permission of Psychonomic Society, Inc.). The finger is shown pushing the right eye to the left through a translation distance *X*. Therefore, in order to maintain fixation on target *A*, the eye must turn through angle Ω, slightly to the right of its initial gaze direction. Furthermore, an exaggerated level of innervation is required in order to rotate the eye against the finger pressure, equivalent to moving the eye through an additional angle *θ* in the direction of gaze represented by the dashed arrow. The left unimpeded covered eye actually turns to that extent. Thus, the behavior of the covered eye is a good indicator of the magnitude of the exaggerated efference necessary to maintain fixation under the eye-press paradigm.

known just what aspects of a structured visual environment control the null effect. Are linear elements crucial? Is density of contour important? Is foveal stimulation as important as peripheral representation of the pattern?

Eye-Press and Altered Vergence Innervation

If, during eye press, binocular viewing of the target is permitted, then proper convergence can be achieved with sufficient innervation to overcome the ocular displacement and finger pressure, as well as any rotary force imparted by the finger. Under these conditions, although vergence and single vision may be maintained, the internally registered innervation is exaggerated or diminished depending on target distance relative to the levels required for convergence when no eye press is attempted. Accordingly, the apparent distance to targets seen under eye press should be distorted in proportion to the level of compensatory vergence innervation required to overcome the eye press. Do these illusions or distortions occur in full illumination? Rine and Skavenski (1997) had subjects view a target seen in a fully illuminated room, and then point at it 1 sec after extinguishing the lights. Pointing measures taken with and without eye press on the left outer canthus showed significant leftward and target overshoot errors due to the eye-press condition. Thus, there is evidence, consistent with Matin et al. (1982) and Stark and Bridgeman (1983), that when perception is measured by pointing, apparent direction as well as distance is governed by ocular muscle innervation even in fully illuminated visual fields.

The Adaptation Paradigm and Altered Vergence Innervation

Another test of the role of ocular muscle innervation in distance perception, in fully structured environments, was provided by Shebilske, Karmiohl, and Proffitt (1983) using an adaptation paradigm. Just as in the case of the paralysis and eye-press paradigms, adaptation causes a change in the level of innervation required to sustain any given ocular posture. This is accomplished in the vergence system by shifting the tonus of the vergence resting level so that after adaptation, innervation levels required for convergence on fixation targets are either lower or higher than they were before adaptation (see Chapter 4). Shebilske et al. (1983) induced changes in vergence resting level and in distance perception by having subjects maintain binocular fixation on a small near target for a 10 min period. Perceived distance was measured via a pointing response with the unseen hand and showed significant shifts of 6.34 cm in darkness and 2.30 cm in a structured, well-lighted room. Similar light–dark differences also were found by Shebilske (1981) after inducing a directional oculomotor bias.

Indeed, a bias induced in the vertical orientation of apparent straight ahead eye position has been shown to alter predictably the swing of a baseball bat in a fully lighted, naturalistic environment (Shebilske, 1986). After playing a game entailing visual and manual skills during which they maintained a downward head pitch, the subjects, who were experienced batters, hit balls pitched from a batting practice machine. The head and eye posture assumed during the game produced a downward shift in ocular straight-ahead resting level, a condition previously shown to increase the apparent height of fixated objects. In comparison to the level of the swing made immediately prior to either the head-upright control or the experimental condition, the number of higher swings relative to lower ones doubled after the head-tilt condition, but stayed about even after the head-upright control condition. There is thus no doubt about the perceptual effects of bias in oculomotor parameters even under complex, full-cue conditions.

Ocular Muscle Vibration

This is a relatively new procedure (see Chapter 4) for biasing the tonus of selected extraocular muscles by applying low-frequency vibration periorbitally (Roll et al., 1991; Velay, Roll, Lennerstrand, & Roll, 1994). When applied (e.g., beneath the eye near the inferior rectus), a fixated target in a dark surround appeared to move upward, and in general opposite in direction to the location of the stimulated muscle. The phenomenon probably results from an induced contraction in the stimulated muscle and the ensuing eye movement (Velay, Allin, & Bouquerel, 1997) that, if opposed by the fixation reflex, would cause some compensating pursuit signal to be issued in a direction opposite the induced reflex. When tested by pointing in darkness the effect was robust, but in a lighted structured surround no illusory movement was detected (Velay et al., 1994).

Although there is thus good supportive evidence to show that oculomotor efference contributes to the perception of direction and distance in the dark as well as in lighted, visually structured environments, several questions nevertheless remain unanswered: Why do target settings to some visual criterion, such as the apparent straight ahead, fail to manifest the perceptual effects of altered oculomotor efference (Matin et al., 1982), and even when pointing is used, why the reduction in measured illusion in light relative to that measured in darkness (Shebilske, 1981; Shebilske et al., 1983)? Likewise, the complete loss of a vibration induced illusion (Velay et al., 1994) in a structured visual environment must be addressed. Two paths to a solution already exist. One, already described, is the suggestion proposed by Bridgeman (1996) that a lighted environment contributes reflexive optostatic control, thereby reducing the level of compensatory or voluntary efference. The second is based on Shebilske's (1987) *Ecological Efference Mediation Theory*, which treats visual and efference-based mechanisms as interacting channels, both of which contribute to perception. Both approaches are promising, but they either need adjustment to accommodate the facts or the facts need refinement. For example, Matin et al. (1983) found no light–dark differences when measured by pointing or auditory matching, but if optostatic stimuli reduce voluntary efference in the light, they should show a reduced illusion, even with a pointing or auditory response measure. Finally, it should be noted that certain mundane factors such as the spatial frequency and orientation of the visual scene pattern as well as the magnitude of voluntary efference brought to bear may play a determining role in any particular result.

As a rule we cannot logically deduce the results of empirical research, which is why empirical science is so useful, but perhaps as a heuristic we may note that there are numerous examples of redundancy in perceptual systems. So, it seems not unlikely that such redundancy is functional. For example, efference-based perception of direction may play a disambiguating role when the full set of determinants of perceptual direction are considered. Further observations on the functional aspects of redundancy in perception are presented in Chapter 6.

Eye Movements and Vection

Introduction

Vection, the act of carrying or state of being carried (Webster's Universal Dictionary, 1936), refers, in the study of perception, to the sense of self-movement induced by exposure to a moving surround. When the illusion is complete, the surround appears perfectly stationary and all movement is attributed to the self. Common examples include the experience of drifting when on a dock firmly attached to land, or the sense of backward movement in a stationary car that impels one to step on the brakes, when actually an adjacent car is moving forward. Vection, of course, no longer is merely left to happenstance but can be produced as desired and may be experienced at theme parks, large screen movies, and in virtual reality devices. It also is used extensively in flight and auto simulators. The experience of self-motion may be absolutely compelling, but what is known of the underlying mechanism?

Oculomotor Mechanism vs. Cognitive State

Ernst Mach (1906/1959), the first to investigate formally the vection phenomenon, proposed that self-motion arose out of the attempt to maintain voluntary fixation while combating the reflexive pull on the eyes produced by ongoing optokinetic stimulation. Figure 5.02 represents one of the pieces of apparatus designed by Mach to study the phenomenon in the laboratory.

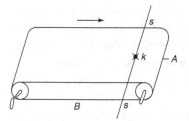

Figure 5.02. *Mach's (1906/1959) Apparatus for the Laboratory Study of Vection.*
(Reprinted from Mach, 1906/1959, *The Analysis of Sensations*, p. 143, S. Waterlow
(Trans.). Reprinted with permission from Dover Publications, Inc.) Mach described
the apparatus as composed of an "oil-cloth of simple pattern... drawn
horizontally over two rollers, two meters long and fixed three meters apart in
bearings and... kept in uniform motion by means of a crank" (p. 143). The knot *k*
in the string *ss* was about 30 cm above the cloth and served as a fixation point for
the observer. With the pattern moving in the direction of the arrow, the observer at
A, and the knot fixated, forward movement or linear vection is experienced. With
the subject at *B*, vection to the left occurs. It seems plausible that the "simple
pattern" produced a low density of contours and that the manually driven
apparatus could be driven only at relatively low speeds. These are the conditions
found by Howard and Howard (1994) to be most conducive to vection, and so
Mach's loss of vection with direct viewing of the pattern is deducible from the
conditions of his study. It certainly is possible, at low stimulus speeds when pursuit
movement is viable, that when fixating on some detail of the pattern, its motion is
sensed and vection is lost. However, under conditions where reflexive OKR is viable,
the oculomotor apparatus does not signal the presence of eye movement or target
motion. In this indirect way, OKR may play a role in the production of vection.

Whenever Mach fixated on the knot above the moving pattern, he experienced
vection opposite the pattern movement along with apparent movement of all
stationary objects in the environment, including the self. However, the illusion
disappeared when the OKR was allowed to occur by changing gaze from the
knot to the moving pattern. As discussed earlier in this chapter, the proposed
underlying mechanism was oculomotor in nature, arising from the substitution of voluntary fixation and pursuit for the reflexive optokinetic response. Is
fixation on a stationary target actually required for vection? The short answer
is no, but it helps (Fischer & Kornmüller, 1930; Howard & Howard, 1994). In a
recent study of circular vection (i.e., apparent yaw movement around a vertical
axis), Howard and Howard (1994) placed subjects within a large vertical cylinder covered over 10 percent of its surface with a randomly placed array of white
spots seen against a black background. The time from the onset of cylinder
motion to the first sense of vection was recorded, and subjects also rated the
extent to which motion was experienced as residing totally in the self, with a
stationary surround, or in some combination of self and surround movement.

Howard and Howard (1994) investigated three cylinder speeds, had a fixation target in one condition and none in the other, and also examined conditions in which different numbers of rods were placed between the observer and the cylinder wall, in comparison with a full field without intervening objects.

It is instructive to consider that in Mach's time such parametric or categorical variation of stimulus conditions was not characteristic of experimentation. Instead, the single-condition demonstration experiment typically was used to support the researcher's point. It is, of course, for all to appreciate that controlled experiments were done at all, so that the move from philosophy to empiricism could get its start. Nevertheless, the inattention to parameters probably was responsible for much unproductive controversy at the turn of the twentieth century. In this connection, note that William James (1890/1950) could not corroborate Mach's observation of full vection while standing on a bridge, with the actually running water appearing to stand still (Mach, 1906/1959, p. 142). In contrast, James (1890/1950) wrote:

> I have myself repeated the observation many times above flowing streams, but have never succeeded in getting the full illusion as described by Mach. I gain a sense of the movement of the bridge and of my own body, but the river never seems absolutely to stop: it still moves in one direction, whilst I float away in the other. (p. 513).

For James, the critical determinant of vection was not ocular innervation, but rather the change in attention such that whatever was regarded as background is perceived as relatively stable, while that which is attended to as object tends to take on the quality of motion. Accordingly, vection represented "the sort of consciousness which we have whenever we are ourselves borne in a vehicle, on horseback, or in a boat. As we and our belongings go one way, the whole background goes the other" (James, 1890/1950, p. 513).

The two approaches of Mach and James, mechanistic and cognitive, respectively, are important sources of competing hypotheses within psychology today. See, for example, Ebenholtz (1990) vs. Rock (1990) on the interpretation of the rod and frame effect; Owens and Leibowitz (1976), Post and Leibowitz (1982), Post and Lott (1992), in contrast with Gogel and Tietz (1974) and Gogel (1973) on the interpretation of apparent motion with concomitant head movement; and Fisher and Ebenholtz (1986) vs. Wallach, Moore, and Davidson (1963) on the adaptability of stereoscopic depth by the Kinetic Depth Effect (KDE). However, as we will see, in the study of vection the two approaches are not mutually exclusive. As for James' view of the critical role of background, it is now well established that the moving scene perceived as background controls the direction of vection, whereas the moving pattern seen as foreground, between observer and background, is without effect. Likewise, when two visual surfaces are present simultaneously, a stationary background prevents vection, whereas a stationary foreground does not (Brandt et al., 1975; Ohmi, Howard, & Landolt, 1987).

There is a phenomenon worth noting in this context that may be characterized as an *antivection phenomenon*. It occurs when approaching port on a large ship or ferry. Here one typically encounters a large land mass welling up in the field of view, and instead of an induced forward movement of the self there is a quite compelling experience of the dock and attached land mass as moving toward the observer, with the observer stable in space along with the ship. The apparent movement of the background would seem to run counter to James' rule. We conclude that although the rule has been supported by experiment, there must remain one or more parameters beyond foreground–background not yet fully explicated that actually control the vection experience.

What about Mach's attempt to isolate an underlying mechanism—namely, his mechanistic hypothesis? First, it seems unlikely that pursuit suppression of the OKR is the cause of vection. Although vection occurs sooner and is rated as stronger when the subject fixates a nonmoving target, it nevertheless occurs even with no fixation target and with the subject undergoing optokinetic nystagmus (Howard & Howard, 1994). It is also true that fixation contributes more to the apparent speed of vection and to the lowering of the vection threshold at slow stimulus velocities than at fast ones (Howard & Howard, 1994). Therefore Mach's report of vection only on fixation of the knot (see Fig. 5.02) may be tied to the actual stimulus velocities used in his studies and also to his particular choice of stimulus pattern, factors now lost in history.

Second, there is evidence that vection and the illusory motion of a fixated target, associated with pursuit suppression of OKN, are influenced by separate variables (Heckmann & Howard, 1991). For example, the illusory movement of a fixated target attributable to pursuit suppression of OKN occurs most strongly when the observer converges on a test target that is in the same depth plane as the optokinetic stimulus itself. Because OKN occurs most strongly when the eyes are converged in the plane of the inducing stimulus (Howard & Gonzalez, 1987), it is to be expected that illusory movement based on OKN suppression also should be influenced by the same circumstance. However, of two concurrently viewed but oppositely moving optokinetic patterns, vection direction is determined by the more distant appearing of the two (Ohmi et al., 1987), regardless of the motion direction of the surface on which the observer actually converges (Heckmann & Howard, 1991). In other words, vection does not require convergence in the plane of the moving stimulus, whereas maximal OKN and OKN suppression do require convergence in the plane of stimulation. Thus, we may conclude that even though the same stimulus pattern of visual stimulation underlies both OKN and vection, they are not causally related. Rather, both vection and OKN may result from the same recurrent optokinetic pattern. This is consistent with the possibility that the two phenomena probably share some portion of the underlying neural circuitry, because the time course of the build-up and dissipation of both follow similar patterns (Howard, 1982, p. 392; Heckman & Howard, 1991). In this respect it has been suggested that visual optokinetic stimulation produces the sense of self-rotation by influencing the same vestibular cells as those actually driven by inertial stimuli (Henn, Young, & Finley, 1974; Dichgans & Brandt, 1978). This parsimonious hypothesis is challenged, however, by the finding that individuals with bilateral labyrinthine defects nevertheless are capable of experiencing vection (Cheung, Howard, Nedzelski, & Landolt, 1989). Apparently a nonlabyrinthine route of stimulation must be available.

Back to eye movements. Any model of vection must offer an account of its two phenomenologic components. First is the experience of a stable, nonmoving scene, and second is the sense of self-movement in space. When OKN occurs, because its slow phase is in the direction of the moving stimuli, it inevitably acts to slow or null out the retinal movement entirely. To a certain extent, therefore, the sense of object motion attributable to image movement across the retina is diminished by the nystagmus itself. This state no doubt contributes to the sense of surround stability. As to the second attribute, that of apparent body motion when eyeballs are rotated reflexively as during VOR or OKN, there is possible logically, an ambiguous interpretation: In the absence of a voluntary pursuit signal, either the eyes may be considered to be moving within the stationary skull or they may be stationary in space with the skull assumed to be moving around the eyeballs. The latter is consistent with vection, and therefore it is potentially informative to consider cases outside of the optokinetic–vection paradigm, in which reflexive eye movements were engendered. Do they produce illusory self movements similar to that characterized by vection?

Ocular-Muscle Vibration, Again

In addition to the apparent movement of a fixated target described by Velay et al. (1994), vibration of selected extraocular muscle also produces postural and kinesthetic illusions (Roll & Roll, 1987). Of particular interest in the context of vection is the report of a slow head rotation that persists as long as the stimulation is applied. Roll and Roll reported that "when the subject was seated with his head free . . . the application of a vibration train to the lateral rectus of one eye, or to homonymous muscles of both eyes, elicited an illusory sensation of head rotation, the direction of which corresponded to a rightward horizontal rotation in the case of left lateral vibrations"(p. 63). Corresponding illusory movements, opposite in direction to the position in which the vibrator was placed, also were obtained in pitch orientation, and, when the head was fixed, "vibration of the same lateral muscles induced kinesthetic trunk illusions in the direction previously described for the head" (p. 64). From these results it seems clear that reflexive contraction of the extraocular muscles is capable of signaling illusory movement of head and trunk. Are there in these results implications for a better understanding of vection? A positive answer hinges on the extent to which OKR and vibration induced oculomotor stimulation share common physiologic features.

Conclusion

Visually induced vection is a functional phenomenon and not just a laboratory curiosity because it probably contributes to the veridical sense of movement when walking or while being transported. To date, scientists have not been successful in isolating any single necessary condition except for the presence of optokinetic stimulation in the form of a moving visual pattern that is registered as background. It remains to be seen whether OKR itself or the internal neurologic signal used to drive the nystagmus actually is the critical feature in vection induction when OKR is suppressed.

A modern vection apparatus is represented in Fig. 5.03.

Eye Movements and Motion Sickness

Stimulation of the vestibular organs of the inner ear leads reliably to eye movements in the form of the vestibular ocular response or nystagmus (see Chapter 3). Accordingly, it is conceivable that the activation of this oculomotor system during exposure to nauseogenic stimulation plays a significant role in the production of motion sickness. A feasible causal nexus has been proposed by Ebenholtz et al. (1994), linking eye movements to many of the vegetative symptoms of motion sickness, and this proposal is examined later. First, however, it is important to describe the phenomenon, note how widespread it is, and consider its possible functional aspects.

Figure 5.03. *A Modern Vection Apparatus.* (From Bles, W. (1979), *Sensory Interactions and Human Posture: An Experimental Study*. Reprinted with permission of the author.) Shown is the apparatus used by Professor Willem Bles at the Free University of Amsterdam. The vertical bar and drum were capable of independent rotation. Bles (1979) studied the interactions of visual, vestibular, and somatosensory factors in the vection phenomenon. It is interesting that vection was reported with the lights out and with the bar stationary but floor rotating so that subjects walked as they would on a circular treadmill. Because they did not rotate in space, the horizontal semicircular canals remained unstimulated, but circular vection occurred nevertheless. Some subjects also exhibited nystagmus (Bles, 1981). Self-motion is thus multiply determined, and even a rotating sound source is capable of inducing both self-motion and nystagmus (Lackner, 1977).

Virtually Ubiquitous. It is a simple matter to find anecdotal evidence of motion-sickness experiences, the provocative causes being so widespread. These include the classic case of a passenger aboard a pitching and rolling ship, but also many less obvious situations, including:

camel and elephant riding,
space travel,
downhill skiing,

auto and airplane simulators,
viewing large-screen videos and movies,
vection-inducing displays,
microfiche reading devices,
wearing newly prescribed eyeglasses,
earthquake movements,
viewing through binoculars,
moving assembly lines,
virtual reality devices.

All of these situations ultimately are capable of producing vomiting or emesis (from the Greek verb *emein*), but they also produce, in varying subsets, a number of other overt signs and subjective symptoms. These include:

eye strain,
headache,
pallor,
sweating,
dryness of mouth,
stomach fullness,
drowsiness and depression,
disorientation and vertigo.

Many years ago, William James (1890/1950) noted that deaf mutes tended not to experience motion sickness and it is now well established that an intact vestibular apparatus is necessary for the occurrence of symptoms, even under such provocative conditions as afforded by a crossing through 40 ft waves during a storm in the North Atlantic (Kennedy, Graybiel, McDonough, & Becksmith, 1968). Fifteen of the twenty control subjects on board vomited at least once, whereas none of the ten labyrinthine defectives did so. For them, drowsiness, subsequently known as the *sopite syndrome* (Graybiel & Knepton, 1976), and fear were the dominant symptoms.

Although the list of stimulating conditions includes many that are strictly visual in nature, like the scenes used with fixed-base flight simulators, this does not mean that the vestibular system is not implicated in these cases, because visual signals are known to influence the very same vestibular-nucleus cells as those that respond to actual inertial stimuli (Waespe & Henn, 1977; Daunton & Thomsen, 1979). Furthermore, those with defective vestibular systems likewise are immune both from visually and inertially induced motion sickness (Cheung, Howard, & Money, 1991). Thus, it is clear that a functioning vestibular system is critical to the process, but before we consider just how a stimulated vestibular system provides for all this misery, it may be instructive to consider what functional value, if any, motion sickness may have.

Functional Significance

a. Treisman's neurotoxin–mimetic theory. Numerous species, including birds, fish, dogs, horses, cows, monkeys, and chimpanzees, exhibit signs of motion sickness (Money, 1990). Furthermore, motion sickness affects not only vegetative body functions but has serious cognitive consequences as well in the form of drowsiness, apathy, and depression (Money, 1970). Is there anything of a positive functional value associated with all these negative signs and symptoms that might have conferred an evolutionary advantage? No scientific consensus has ever formed an answer to this question, but an ingenious proposal has been formulated by Michel Treisman (1977) at Oxford University. Treisman's hypothesis is based on the very plausible observation that there are many circumstances in which vomiting is good for one in the sense that it helps rid the gut of ingested toxins. Because emesis is a highly functional response to stomach poisoning, then if the body's response to motion sickness–inducing conditions could be shown to be equivalent to its response to poisoning, motion sickness would become understandable, although still not in itself functional. To complete the case, Treisman borrowed heavily from the sensory conflict theory of motion sickness (Reason & Brand, 1975) by suggesting that motion sickness arises when "an individual is placed in a situation in which . . . one type of input is repeatedly misleading in what it predicts for another, and this affects some skilled motor performance such as fixation or maintaining head position" (Treisman, 1977, p. 494). In other words, consider an observer in a closed cabin on a rolling and pitching ship. Here the visual scene presumably signals no relative movement, therefore a stable environment, whereas the vestibular system is in conflict by registering the true inertial state of the ship. Accordingly, Treisman speculated that neurotoxins and our response to motion sickness stimulation both entail alterations "in sensory inputs or motor coordination or both. Even a minor degree of impairment of sensory (vestibular) input or of the coordination of eye muscles would produce mismatches between the systems. An emetic response to repeated such mismatches would be an advantageous adaptation" (p. 494). This would be so after ingesting neurotoxins, but it would be an inappropriate although inevitable response to the same mismatches generated by motion sickness stimulation. In short, motion sickness occurs when the conditions mimic the sensory conflicts that normally are produced by ingested neurotoxins.

Even if it is accepted that neurotoxins may produce conflicts among signals regulating visual, vestibular, and proprioceptive systems, and that similar conflicts occur under motion sickness–stimulating conditions, there is no independent evidence that sensory conflicts cause emesis. Nor is there an underlying logically drawn rationale as to why this should be the case, or why the many other signs and symptoms of motion sickness should be triggered by sensory conflicts. Nevertheless, Treisman's theory remains as an ingenious and unique attempt to find evolutionary utility in the motion sickness syndrome.

Before leaving Treisman's evolutionary hypothesis, it is important to note that an indirect test exists of Treisman's speculation that conflicting sensory signals, especially those from the vestibular system, are a cause of emesis. Consistent with this view, Money and Cheung (1983) found a significant reduction of the emetic response to certain poisons in seven dogs, each of which had had the inner ear surgically removed. Because the absence of the vestibular organs reduced the likelihood of vomiting, Money and Cheung (1983) concluded that one of the functions of the inner ear was to facilitate the emetic response to poisons and that, consistent with Treisman, motion sickness arises when this same mechanism is triggered inadvertently by motion sickness–inducing stimulation. It seems clear from the research of Money and Cheung (1983) that the inner ear plays a significant role in the emetic response to certain poisons. It does not follow, however, that conflicting visual, vestibular, and proprioceptive signals also mediate the emetic response. What then, might the underlying mechanism be?

b. Positional alcohol nystagmus (PAN): A better model for the inner ear facilitation of the emetic response to poisons.

It is interesting to observe in this context that Money and Myles (1974) actually explicated the physiologic mechanism responsible for the nystagmus and the motion sickness syndrome, produced by heavy water (deuterium oxide) and also ethyl alcohol ingestion. These poisons produce a phenomenon termed *positional alcohol nystagmus* (PAN), in which nystagmus and the motion sickness syndrome are triggered when the subject assumes certain head positions in relation to gravity. For example, after considerable alcohol ingestion, and with the head held right-ear down, the nystagmus occurs with fast phase to the right, slow phase leftward, and the opposite relationship is obtained when the head is placed left-ear downward. The phenomenon was regarded as in need of explanation because nystagmus normally is governed by the cupula of the semicircular canals (see Chapter 3), which does not respond to the linear acceleration of gravity, but does respond to radial or angular acceleration. Why should a static change in the direction of gravity influence the canals after ingesting ethyl alcohol? Money found that, after ingestion of deterium oxide, PAN occurred, but with a direction exactly opposite that produced by ethyl alcohol. He reasoned that normally, the cupula floats at neutral buoyancy in the surrounding endolymph within the canals (see Fig. 3.15c). Therefore it would not be disturbed by changes in the gravity vector when tilting the head. However, with absorption of heavy water into the blood stream, the cupula may be expected to lose its neutral buoyancy, becoming heavier than the surrounding contaminated endolymph. With alcohol ingestion, just the opposite response is expected because alcohol is lighter than the endolymphatic fluid. Therefore the cupula would float or move with the surrounding fluid when absorbing alcohol, but sink or move against the surrounding endolymphatic stream after absorbing deuterium oxide. These opposed actions thoroughly account for the opposite kinematics of the nystagmus that occurs after ingesting the two poisons. Money and Myles (1974) clinched their argument by showing, in the cat, that ingesting both alcohol and heavy water together actually neutralizes the nystagmus that would otherwise be present as a result of drinking either chemical by itself.

It is clear from Money's research that the two poisons, deuterium oxide and ethyl alcohol, wind up being sampled in the cupula and, by changing the relative specific gravity of cupula and surrounding endolymph stimulate the canals to produce nystagmus accompanied by motion sickness. In light of these facts it seems reasonable to presume that other poisons, including possibly those given to the dogs in the Money and Cheung (1983) experiment, also will be sampled at the cupula. For those poisons, the emetic response to which is facilitated by the semicircular canals, perhaps the underlying mechanism simply is the relative change in specific gravity of cupula and endolymph. Accordingly, perturbation of visual, vestibular, and proprioceptive inputs, and attendant sensory conflicts, may be quite irrelevant. Consistent with the core idea of Treisman's theory, the functional significance of motion sickness may lie simply in the limited functionality afforded by the absorption of certain poisons in the cupula and endolymph of the inner ear and the subsequent stomach emptying brought on by emesis.

We turn now to a possible oculomotor link between vestibular stimulation and motion sickness.

Eye Movements and the Vagus Nerve Connection

The nuclei of the vagus, the tenth cranial nerve, are situated in the brain stem (Keele & Neil, 1965) and form a major component of the autonomic nervous system. The name *vagus* refers to the vagrant or wandering aspect of the nerve that supplies many of the internal organs. For example, in the heart all activities are depressed by the vagus nerve stimulation, and vagal influence has been found in the release of insulin, in the tone of stomach and intestinal muscle, in esophageal peristalsis, in breathing rhythm, and in the contraction of the walls of arterioles and veins (Keele & Neil, 1965). This is important for an understanding of motion sickness because many of its signs and symptoms seem to be triggered by vagal activity. In fact, recent research has provided neuroanatomic evidence for a linkage between vestibular nucleus sites and brain stem regions, including the dorsal motor nucleus of the vagus nerve (Porter & Balaban, 1997). Therefore, it is becoming possible to describe the neural mechanisms that underlie provocative stimulation of the vestibular system and the various vegetative signs and symptoms of motion sickness that ensue.

But what of visually induced motion sickness produced by a wide-screen movie or a vertically striped rotating drum (Kennedy et al., 1996; Hu et al., 1999), similar to that shown in Fig. 5.03? Because optokinetic stimulation is known to modulate vestibular nucleus cells (Waespe & Henn, 1977), the same mechanism may be assumed to underlie both inertial and visually stimulated sickness. It may be significant that in both cases, the presence of nystagmus provides the potential for an added feature, namely, an ocular-muscle contribution to motion sickness. This hypothesis was proposed by Ebenholtz et al. (1994) based on two sources of evidence. First was the *oculocardiac reflex*, a phenomenon in which cardiac activity may be diminished to the point of arrest as a result of a rapid stretching or traction of extraocular muscle (Milot, Jacob, Blanc, & Hardy, 1983). The reflex is especially likely during strabismus surgery, where one or more extraocular muscles must be lifted in preparation for surgical shortening. The neural pathway has been described as running from the ciliary ganglion to the trigeminal and vagus nerves. (Katz & Bigger, 1970). Thus afferent signals from rapidly stretched extraocular muscles ultimately stimulate the vagus nerve. The second basis for the hypothesis was the finding that anesthesia applied to the eye muscles behind the eye produced a significant reduction in the incidence of emesis and nausea after strabismus surgery (Houchin, Dunbar, & Lingua, 1992). Presumably, the application of retrobulbar anesthesia, by blocking the afferent signals emanating from stretched extraocular muscle, was responsible for reducing the level of emesis to that occurring after general anesthesia in nonstrabismus surgery. To summarize: traction, or rapid stretching, causes afferent signals from extraocular muscle to stimulate vagal activity, and inhibiting these signals, via anesthesia, reduces the nauseogenic effects of vagal activity. Do these conditions apply under more typical eye movement situations? Does eye-muscle traction ever occur as a result of the operation of the "natural" eye movement control systems?

a. The Coriolis maneuver and eye-muscle traction. The answers are a matter of conjecture, largely because neither eye movements nor eye muscle tension have been recorded under conditions that might be thought to produce eye-muscle traction. However, a logical argument may be made in support of "naturally" occurring traction. Consider, for example, the Coriolis maneuver, a procedure frequently used to induce motion sickness in the laboratory. A typical procedure is represented in Fig. 5.04, where the rotating subject is shown pitching his head forward, before bringing it back to the upright position. The Coriolis maneuver, also known as a *cross-coupled rotation*, is an exceedingly effective inducer of motion sickness for reasons that are not now understood, but the effect of this maneuver on the semicircular canals may suggest a reason. Figure. 5.05 is a geometric representation of one set of canals (see Chapter 3, Fig. 3.15) with head upright (a) and after a 90-deg shift in head position (b). The maneuver is always performed after the subject has been rotating sufficiently to allow the cupula of the horizontal canal to settle into its neutral resting state. This usually occurs within about 25 sec of the initial acceleration, at which point the endolymph in the canal is moving with the same velocity as the surrounding skull. In the dark, when rotating at constant velocity, there would be no clue from the canals of any body rotation whatsoever. When, as shown in Fig. 5.05(b), the subject pitches the head forward, one of the vertical canals moves into the horizontal plane, where its endolymph begins to rotate until it catches up with the rotational movement of the skull. Contrariwise, the horizontal canal moves into a vertical plane, where its endolymph undergoes a deceleration because no rotational stimulus is received in this plane. In point of fact, all three canals undergo some activity because the canals are not perfectly orthogonal, nor is the head movement perfectly executed in a vertical plane. The net effect, a vector sum of each canal's stimulation, is to induce a roll sensation of falling to the left or right, depending on the direction of rotation about the vertical axis, counter-clockwise or clockwise, respectively (Howard, 1982, pp. 347–348). Likewise, if a head movement towards one shoulder is used as the Coriolis maneuver, the resultant sensation is an apparent pitching of the subject. In either case, the experience is of an apparent rotation about an axis perpendicular both to the vertical axis of body rotation and the axis of head rotation (Lackner, 1978).

Yet, what about oculomotor factors such as nystagmus and torsion? There is no doubt that nystagmus is generated by the Coriolis maneuver. For example, Melvill Jones (1970) found an oblique nystagmus, 45 deg off the horizontal,

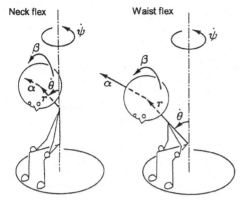

Figure 5.04. *A Typical Coriolis Maneuver.* (Adapted from Isu, Yanagihara, Mikuni, and Koo, 1994, Coriolis effects are principally caused by gyroscopic angular acceleration, *Aviation, Space, and Environmental Medicine, 65,* 627–631. Reprinted with permission of Aerospace Medical Association.) Coriolis effects were worked out by the French mathematician Gustave-Gaspard de Coriolis (1835). They typically refer to the distribution of forces apparently acting on an object in linear translation relative to a rotating system. For example, in attempting to walk on a radial line from the edge to the center of a rotating merry-go-round, the walker would be moved off track as if there was a force acting in the direction of the merry-go-round rotation perpendicular both to the merry-go-round axis of rotation and the walkers intended path. Thus, if the merry-go-round were moving counterclockwise as seen from above, the Coriolis force would deviate the walker to the right. This is understandable, considering that the rightward speed of the platform diminishes to zero as the walker approaches the center. So with each step the walker will be carried slightly too far to the right with respect to the intended radial path. Similar considerations apply to a rocket fired from the equator toward the North. The rocket will leave the equator with an eastward speed of about 1000 miles per hour due to the Earth's rotation, but it will land to the north of the equator where the surface of the earth may be moving at only about 850 miles per hour as at the New York latitudes. Consequently, the rocket will land too far to the east, simply because the Earth's speed at the landing site is slower than that at its take-off point. The rocket will behave as if a force, the Coriolis force, had acted to move it eastward.

In the figure, $\dot{\psi}$ represents angular velocity of body rotation, $\dot{\theta}$ is the angular velocity of head pitch, and r is the radius of head rotation centered either at the neck on the left or waist on the right. β represents a spiral or gyroscopic angular acceleration produced by the vector sum of $\dot{\psi}$ and $\dot{\theta}$.

By moving the head forward while seated on a rotating platform, the two figures are subject to Coriolis forces α, acting to deviate the head to the right. This is stronger in the case where movement occurs from the waist, as in the right figure, instead of from the neck, because the head departs more so from the axis of rotation where head velocity is low. Isu et al. (1996) found that although the waist-flexing condition produced about seven times the strength of Coriolis force, both conditions were equally effective in causing nausea. Therefore Coriolis forces probably are not the causal agents for this type of motion sickness. Isu et al. (1996) did find, however, that nauseogenic effects varied with the magnitude of the gyroscopic angular acceleration.

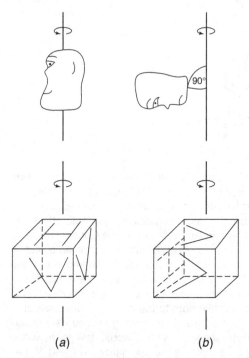

Figure 5.05. *Cross-Coupled Rotation.* In the Coriolis maneuver, the subject is rotated at constant velocity around a vertical axis. When the head is rolled or pitched, as in this figure, head rotation then occurs simultaneously around two axes. (a) The stylized horizontal (*H*) and vertical (*V*) canals in their respective planes. (b) The changed orientation of the canals after a 90-deg head pitch. In the figure, a vertical canal whose endolymph previously was inert reacts when it enters the horizontal plane of rotation, whereas the endolymph in the horizontal canal gradually decelerates because it is taken out of the plane of rotation. Some activity also occurs in the third vertical canal because the canals are not perfectly parallel with the horizontal and vertical planes of space. A complex nystagmus is initiated by the vector sum of canal activity and, in addition to feeling sick, the subject senses head movement in a plane orthogonal to the two planes in which the subject moved. For the rotations shown in (b), the sensation would be that of leftward rolling (Isu et al., 1994).

after a Coriolis maneuver to the left shoulder, and although data are lacking, it is reasonable to assume that the otolith–ocular counterrolling reflex occurred concurrently. Imagine an eyeball rotated around its line of sight while simultaneously undergoing a reflexive oblique nystagmus. Clearly, the grounds are present for eye muscle traction to occur under these circumstances because the torsional response will surely shorten some muscles also involved in the nystagmus reflex. In any event, little is known of the operation of Sherrington's law of reciprocal innervation (see Chapter 3) when applied to compound reflexive eye movements, where dyschronicities in reciprocal innervation appear likely. It may be concluded that traction events and consequent vagus nerve stimulation certainly cannot be ruled out as a mechanism contributing to motion sickness whenever oculomotor reflexes are triggered, especially nystagmus.

b. Supporting evidence. On the question of empirical evidence, it is worth noting that the suppression of OKN, by providing a fixation point, significantly reduced symptoms of motion sickness induced by viewing a striped rotating drum (Stern et al., 1990). Furthermore, conditions that optimized the frequency of nystagmus beats, in response to the number of stripes on a rotating drum, were found to "accompany the most severe symptoms of motion sickness" (Hu et al., 1997, p. 310). Likewise, in support of the nystagmus-traction hypothesis, on the premise that the higher the frequency and/or velocity of nystagmus, the greater the likelihood of traction events, are two additional sets of results. These are (1) the finding of a higher VOR gain in motion-sickness susceptible, as opposed to nonsusceptible, subjects (Gordon et al., 1996); and (2) the reduction in gain of both vestibular and OKN produced by the antimotion sickness drugs scopolamine and dimenhydrinate (Pyykko, Schalen, & Matsuoka, 1985). There is thus some accumulating evidence for a causal role for nystagmus and indirectly for the traction mechanism in motion sickness. The question raised by Hu et al. (1997), whether nystagmus of the nonoptokinetic variety would likewise modulate the symptoms of motion sickness, is a good one demonstrating the everlasting need for continued research.

In this chapter, the attempt has been made to represent a sample of the broad array of perceptual phenomena to which oculomotor systems may contribute. There are many intriguing problems spawned by this approach, some of which are framed in Chapter 6.

6

Concluding Remarks

Summary

I have sought to describe the many functional properties of oculomotor systems and also to show that among them, several contribute to the spatial qualities of experience: convergence and accommodation signal distance in near space (Fisher & Ciuffreda, 1988; Mon-Williams & Tresilian, 1999), and also modulate the sense of depth and size (van Damme & Brenner, 1997); saccades and the fixation reflex they subserve signal the direction of a target from the egocenter (Matin & Pearce, 1965; Lewald & Ehrenstein, 2000). Not quite so obvious is the fact that this directional information also controls other perceptual qualities such as apparent slant and inclination of surfaces as well (Ebenholtz & Paap, 1973, 1976; Backus, Banks, van Ee, & Crowell, 1999).

On the dynamic side, evidence has shown that the slow pursuit system imparts the quality of motion and length of movement path to fixated objects (Mack & Herman, 1972; Honda, 1990) regardless of whether they actually are in movement: when engendered as compensation for optokinetic nystagmus (OKN) or the vestibulo-ocular reflex (VOR), the pursuit system imbues a fixated and stationary object with illusory movement (Mach, 1906/1959; Post & Leibowitz, 1985). Also to be noted is that, from a methodologic perspective, studies have been emphasized that have demonstrated changes in space perception as a result of modulation of the control parameters or the states of various oculomotor systems (e.g., Gauthier & Robinson, 1975; Ebenholtz & Fisher, 1982; Mon-Williams & Tresilian, 1999).

Implications

What are the implications of failing to recognize the perceptual contributions of oculomotor systems? The foremost and simplest answer is that it would lead to incomplete knowledge of our visual and perceptual systems. This in turn would encourage an undue emphasis on theories of perception that attempt to extract perceptual qualities solely from information in the optic array (Gibson, 1979; Cutting, 1986) or that derive perception from cognition (Wallach, 1976; Rock, 1983). Perhaps most importantly, however, is that a wonderful set of problems associated with the perceptual function of oculomotor systems in particular and consciousness in general would remain unformulated, and therefore, unresolved. Several of these are described in the following.

Unresolved Issues

There are, of course, many problems that remain to be investigated. Following are just a few whose solution would seem to have rather broad implications.

The Perceptual-Integration Problem

The perceptual attributes our oculomotor systems enable are not uniquely motor driven. Rather, many perceptual qualities, including motion and distance, are multiply determined, thereby endowing us with redundant systems. For example, motion is perceived when a contrasting edge moves across the retinal surface as well as when an edge is tracked in the pursuit-movement mode; relative depth is experienced from binocular stereopsis as well as from motion parallax (Bradshaw & Rogers, 1996); and apparent slant about a vertical axis is determined by the direction of gaze (Ebenholtz & Paap, 1973, 1976; Backus et al., 1999), stereopsis, texture, and perspective (Allison & Howard, 2000).

It would appear to be the case that there is great functionality associated with redundant systems; namely, in that they provide the capability to disambiguate otherwise ambiguous percepts. For example, motion parallax without correlated head movement is subject to depth reversals (Braunstein, 1976; Rogers & Graham, 1982), whereas stereopsis faithfully signals the depth sign. Therefore, it is likely that the cooccurrence of multiple perceptual cues provides the occasion for a great reduction of uncertainty. Arguing to a similar point, Shebilske

(1987) suggested that "the integration or synthesis of efference-based and light-based information is more reliable than either source alone" (p. 207). Another aspect of the perceptual integration issue arises when we consider that although parsimony suggests that the same brain activity underlies all instances of the same experience, how to conceive of this in neurologic terms is not presently understood. This problem is especially challenging when the experience in question may be produced by ostensibly different routes of stimulation.

In addition to the reliability-benefits of redundancy, other forms of complementarity also are possible between perceptual states that derive from oculomotor systems on the one hand and, on the other, those based exclusively on retinal stimulation. For example, acceleration in darkness around one of the three spatial axes stimulates the VOR and also produces a sense of body rotation. But this experience gradually diminishes if acceleration ceases and movement continues at a constant velocity, in which interval all sense of movement soon disappears (Israel et al., 1995). If, however, eyes are opened and a patterned environment is present, the sense of movement may continue unimpaired. Likewise, when a near-convergence posture is maintained in darkness, a loss of the sense of convergence magnitude ensues after about 2 min (Ebenholtz & Citek, 1995; see also Fig. 3.10 and the discussion of sentience in Chapter 5). In this case, it also seems likely that with pattern vision available, subjects would then continue to sense their near vergence posture. We conclude that steady-state postures that have been presumed to signal gaze location (Ebenholtz, 1986) may do so only over short, sustained intervals beyond which retinal inputs may be necessary to provide location information. Thus, complementarity of function represents one solution to the question of the "essential coordination and functional interdependence of the higher-order optical and oculomotor information" (Owens & Reed, 1994, p. 272).

The Vection Problem

The role that an oculomotor system such as OKN plays in the vection experience is not understood. Is explicit nystagmus or an implicit nystagmus signal always necessary to stimulate vection, or can vection be produced in the complete absence of oculomotor activity simply by an appropriate motion signal in cortical pathways? Although Roll and Roll (1987) provided evidence for an illusion of head rotation induced by external eye muscle stimulation, it remains to be determined whether reflexive eye muscle stimulation produced internally through OKN is the mechanism underlying vection.

The Sentience Problem

Why are certain oculomotor systems such as OKN and VOR without conscious correlates, whereas others, such as pursuit and voluntary saccades, have strong sentient properties? The gamma-efferent system is known to modulate muscle-spindle output (Whitteridge, 1959). If voluntary behavior is produced by the activation of the gamma-efferent system (Merton, 1972) and if sentience is correlated with increased muscle spindle afference (Matin, 1976; Ebenholtz, 1986), then the link between the gamma-efferent system, voluntariness, and sentience may provide an important path toward a solution.

The Mind–Brain Problem

This is perhaps the most far-reaching problem of all, namely, the problem of the nature of the neural networks that underlie the experiences mediated by oculomotor systems and, in general, the problem of how to characterize the nature of brain architecture and neural activity that provides for conscious and nonconscious states. In the Introduction to this book, I alluded to emergence theory (Scott, 1995) as one of several views of the mind–brain problem. If consciousness is considered to be a special attribute of brains, and indeed of only certain parts of brains, then treating consciousness as an emergent property of an activated brain would seem to be a compelling heuristic to follow. The further question of whether to assign causal properties to conscious states, as Sperry (1985) has advocated, also remains worthy of consideration. Whenever we encounter statements of causal relationships, where it is claimed that perceptions cause perceptions, as between, for example, apparent elevation or displacement and apparent slant or between apparent distance and apparent size or depth, we might consider whether these ultimately lead to tautologic explanations (Kaufman, 1974; Kaufman & Kaufman, 2000, p. 505), and whether problems of this type may be more fruitfully reformulated in terms of underlying cues or stimulating conditions. In support of the latter, see Li and Matin (1998) for an example of the failure of visually perceived pitch to influence visually perceived eye level [see also Fig. 4.07(a–c)], or Ebenholtz (1977b) for an example of the failure of perceived frame size to influence the apparent vertical, although *retinal* size has great influence. Of course, there may be a separate domain of visual phenomena where perceptual interactions are in evidence, as in the case of the effect of perceived spatial layout on apparent lightness (Gilchrist, 1977) or of figure–ground reversals on apparent contrast (Coren, 1969).

It seems likely that two other nonoculomotor problems also would be solved when the brain-architecture problem is explicated. These are: (1) the *locus problem* whereby certain experiences convey an outward or externalized locus while others are centered on or within the observer, and (2) the *modalities problem* whereby certain experiences have distinctive qualities such as the differences between sight, sound, and olfaction. One can only look forward in awe to this kind of enlightenment.

Those of you who do not know the torment of the unknown cannot have the joy of discovery which is certainly the liveliest that the mind can ever feel. But by a whim of our nature, the joy of discovery, so sought and hoped for, vanishes as soon as found. It is but a flash whose gleam discovers for us fresh horizons, toward which our insatiate curiosity repairs with still more ardor. Thus even in science itself, the known loses its attraction, while the unknown is always full of charm. (Claude Bernard, 1865/1957).

APPENDIX

The Ametropias and Other Common Visual Anomalies

Introduction

The term *ametropia*, in Greek means an irregular eye. It is taken to mean an abnormal refractive condition and may be contrasted with *emmetropia*, the normal refractive state. Although emmetropes need no corrective lenses to see clearly at near and far distances, they may suffer from other visual disorders not directly related to focusing, but associated, for example, with the control of eye position or eye movements.

In general, visual anomalies occur when some part of the visual system departs from optimal function. This can occur in a very large, although finite, set of ways. For example, the visuomotor control of the two eyes may be inadequate, causing uncoordinated eye movements; the light gathering and focusing elements, such as the cornea and lens, may not have the proper curvature or the eyeball size itself may not be matched to the focusing mechanism, so that light is not brought to a sharp focus at the retina; biochemical inadequacies may cause light transduction at the photoreceptors (rods and cones) to fail so that electrical stimulation of the neural pathways in the brain is compromised, leading to a loss in detection of fine detail or colors or objects in peripheral vision. Furthermore, the neural projection system from retina to brain may be inadequate, thereby leading to problems in object localization. Defects in virtually any portion of the system can and do occur to create thousands of visual disorders.

Here, eleven very common visual disorders are introduced and described, and clues to the underlying causal mechanisms also are traced. Visual anomalies discussed include:

Myopia
Hyperopia
Presbyopia
Astigmatism
Anisometropia
Aniseikonia
Amblyopia
Strabismus
Glaucoma
Heterophoria
Asthenopia.

Aside from the opportunity to learn a bit of the Greek language, the study of dysfunctional visual states also presents the occasion to gain insight into the normal visual mechanisms that subserve visual perception.

Myopia

The term *myopia* arises from the combination of the Greek verb *myein*, to close, and *ops*, the eye, perhaps because uncorrected myopes would tend to blink or squint.

Myopia is the condition described as nearsightedness because in order to detect small detail, the object must be brought near to the observer. Figure A.00 shows why this is the case. Parallel light is refracted at the second principal plane H_2 to a point F that is short of the retina. However, if light from a nearby target strikes the hypothetical refracting surface H_1, the same refracting power, now added to a diverging beam, moves the focus toward the retina. This is represented in Fig. A.01, where a target point P is shown at the far point, the farthest point from the eye that permits the light to be sharply focused on the retina. Any negative lens that adds the same degree of divergence to parallel light as that obtained if the light actually were emanating from the far point, also causes the myopic focus to shift toward the retina. Therefore, the typical amelioration of myopia is by a properly chosen negative lens.

Like many other dysfunctional states, myopia actually represents a class of disorders, the most common type of which is juvenile onset myopia, typically occurring after the age of five. An adult onset variety occurring after puberty has been recognized by a National Research Council (1989) working group and is a phenomenon many college students have experienced first hand.

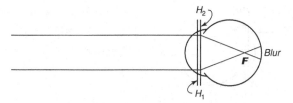

Figure A.00. *Excessive Refraction in the Myopic Eye.* Parallel light from a point on a distant object is represented as being refracted at the second principal plane H_2 of the eye. In the myopic eye this brings the light rays to a focus F in front of the retina, the point being spread over a blurry region on the retina.

In a commonly used visual acuity test, sensitive to near and far sightedness, patients are shown an eye chart on which letters have been printed in various thicknesses and gap sizes so as to project to the eye details ranging from 0.5 to 10 min of arc on different lines. When using such a chart, the designation 20/20 means that at the standard distance of 20 ft, the patient, by identifying the letters, could discriminate a detail that projects 1 min or arc at 20 ft. Likewise, 20/40 acuity means that at 20 ft, the patient could discern detail that project 1 min arc at 40 ft, or equivalently, 2 min arc at 20 ft. With acuity of 20/200, where detail projecting 1 min of arc at 200 ft, or 10 min of arc at 20 ft, is required for letter discrimination, the commonly applied criterion for *legal blindness* has been reached. Note that as the blur, represented in Figs. A.00 and A.02, increases in size, larger letter detail and gap sizes are needed for discrimination.

Currently, the metric designation 6 m is replacing the 20 ft standard.

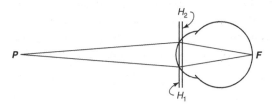

Figure A.01. *Diverging Lens Power Corrects the Myopic Eye.* The total refracting power of the eye represented at the first principal plane H_1 is now added to a diverging beam, as opposed to the parallel beam of Fig. A.00. As a result, the light is converged to a point closer to the retina. If P is at the far point, then the image of P is focused on the retina at F.

A report of the National Advisory Eye Council of the National Eye Institute (Vision research: A national plan: 1994–1998) notes that, "By adulthood, about 25 percent of all Americans are myopic" and that "Refractive eye examinations cost consumers $1 billion annually, and more than $1.5 billion is spent each year on eyeglasses" (pp. 247–248). Contemporary treatment approaches, whose costs have not been estimated because they are relatively new, include radial keratotomy and photorefractive surgery. Both of these procedures

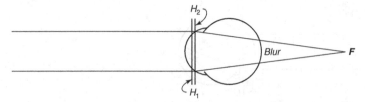

Figure A.02. *Refraction in the Hyperopic Eye.* Parallel light rays from a point on some distant target are focused behind the retina in the hyperopic eye. Thus a single point on the object is represented as a blur region at the retina. Because points on a near object would present the principal plane H_1 with a diverging bundle the focal point in that case would be even farther behind the retina.

entail flattening the corneal surface so that light from distant targets undergoes lower degrees of refraction at the cornea and is therefore focused at the retinal surface rather than in front of it.

Hyperopia

A hyperope is described as farsighted because targets at a distance are easier to resolve than are nearer objects. Unlike the myope, there is no distance at which the image of a target detail is in focus unless the hyperope accommodates or increases the refracting power of his lens. This is demonstrated in Fig. A.02 for a target at optical infinity. To avoid the necessity of constantly accommodating, and the problems that result from this, a positive lens, that would add vergence power to the parallel light bundle before it is processed at the principal plane, is the typical solution to the hyperopic state.

Refractive Error and Axial Length Distributions

We have seen that in order to sharply focus light from a distant source, myopes require negative lenses whereas hyperopes must add positive lenses. The precise value of the correction may be regarded as a measure of the refractive error. When the refractive error is within about $\pm 0.5D$ of 0, the person is classified as an emmetrope whose refractive error falls between that of the myope and hyperope. When regarded in this fashion, the refractive error categories may be treated as regions on a continuum, ranging from high minus through zero to high positive values. In general, studies of refractive error in large populations fail to show a normal distribution. Instead, a typical finding has been that represented in Fig. A.03 from a study of 1,000 right eyes

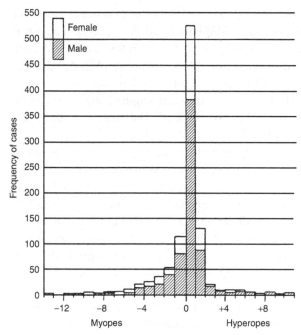

Figure A.03. *Distribution of Refractive Errors of 1000 Eyes.* (Adapted from S. Stenstrom, 1948, Investigation of the variation and the correlation of the optical elements of human eyes. D. Woolf (Trans.), *American Journal of Optometry and Archives of American Academy of Optometry*, Monograph No. 58, 1–71 © The American Academy of Optometry, 1948.) Data are the frequencies of refractive errors in the right eye found in a population of 1,000 clinic patients, army and air corps officer candidates, and physicians and nurses at the ophthalmologic clinic, University of Upsala, Sweden. The ordinate represents the number of cases, and the abscissa distributes the data into 1-diopter bins. More than 50 percent of the eyes had a refractive error between 0 and 1 diopter and would be regarded as emmetropic. This is a significant excess over what would be found if refractive error were normally distributed into a Bell curve. Note also the slight skew to the left because of the increased numbers of myopic eyes.

by Solve Stenstrom (1948). Two important characteristics of the distribution are its asymmetry or skewness and leptokurtosis or peakedness. The higher number of subjects with negative refractive errors (myopes) over those with positive errors (hyperopes) is a rather typical outcome of such demographic studies, as is the large proportion of cases at the 0 to +1 bin. Thus, there are many more emmetropes and near hyperopes than would be expected if the refractive errors were normally distributed to form a typical Bell curve.

Just what the cause of the distribution of refractive errors is remains a source of speculation and hypothecation. After examining correlations between refractive error and a number of dioptric components, such as lens power and corneal curvature, Stenstrom (1948) concluded that "the axial length is the most important of those quantities determining the refractive error" (p. 69). Actually, axial length, the distribution for which is shown in Fig. A.04, correlated to the extent of −0.76 with refractive error, thereby accounting for 58 percent of the variance. Thus, the longer the axial length of the eye the more likely one is to be myopic, while the shorter the eye, the greater the likelihood of finding a hyperope.

Before leaving this topic it is important to recall that correlations do not of themselves indicate causality. Thus, it is just as possible that long eyes cause myopia as it is that myopia causes axial elongation and that some underlying process may cause both myopia and elongation. Extensive near work entailing careful observations at close distances has been thought for centuries to cause some forms of myopia (Curtin, 1985). Certainly the occurrence of myopia concurrently with elongation in adult populations (McBrien & Millodot, 1987), the late-onset variety, suggests that some set of experiential factors associated with near work may be at play (Young, 1963; Ebenholtz, 1983). So pervasive is axial elongation among myopes, that whatever the underlying mechanism turns out to be, it must account for axial elongation. One such potential mechanism, accommodative hysteresis, is discussed briefly in Chapter 3.

Presbyopia

Presbyopia literally means "ancient eye" and is a condition of the aging eye in which the nearest point that can be sharply focused approaches the farthest point that, likewise, can be sharply focused. In the limiting case, only one point at some fixed distance from the observer can be seen sharply. Objects nearer than this point may require a positive lens, whereas objects beyond the critical point typically require a negative lens; therefore the need for bifocals or multifocal spectacles.

Lest one think of this condition as associated exclusively with grandparents, we note that the presbyopic process starts at about 18 years of age with a recession of the near point. This, presumably, reflects the progressive hardening of the cells of the human lens, rather than a loss of the ability of the ciliary muscle to contract (Semmlow, Stark, Vandepol, & Nguyen, 1991; Glasser & Campbell, 1998). Accordingly, the lens loses its elasticity and can neither assume a maximally thin low curvature state for distant objects nor a maximally thick, high curvature state for near objects.

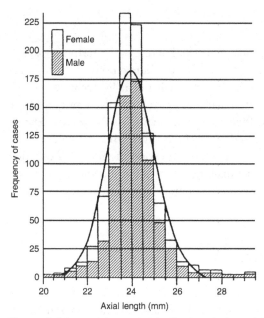

Figure A.04. *Axial Length Distribution*. (Adapted from S. Stenstrom, 1948, Investigation of the variation and the correlation of the optical elements of human eyes. D. Woolf (Trans.), *American Journal of Optometry and Archives of American Academy of Optometry*, Monograph No. 58, 1–71, © The American Academy of Optometry, 1948.) It is of interest to note that Stenstrom's method for the measurement of the axial length of the eye, the distance from the corneal vertex to the retina, was rarely used and by contemporary standards probably would be regarded as imposing an unnecessary health risk. The technique is based on the fact that in the dark-adapted eye, the photoreceptors of the retina are sensitive to x-ray radiation. Stentstrom wrote that:

> The principle of the method lies in directing a pencil of x-rays into the eye in a plane perpendicular to the optical axis. The x-rays cause an excitation of the retina of the dark adapted eye at their circular intersection, so that a dimly luminous ring is perceived. When the pencil of x-rays is shifted backwards, the perceived ring shrinks to a luminous point which occurs when the pencil reaches the posterior pole. It disappears completely when the pencil of x-rays falls behind the retina. (Stenstrom, 1948, p. 18).

By measuring the position of the corneal surface and the posterior pole of the eye on calibrated scales, the total axial length was determined.

The curve is the normal function associated with the total set of 1,000 eyes. The distribution shows slight skew toward longer eyeballs and also is leptokurtic because of the excess of eyes at about 24 mm. The highest frequencies found at bins for 23.5–24.0 mm and 24.0–24.5 mm correspond to the emmetropic overpopulation at the center of Fig. A.03 and is consistent with the finding of a significant correlation between axial length and refractive error, the longer eyes being associated with greater myopia.

Astigmatism

Stigma is the Greek word for point or mark. Therefore the term *astigmatism* refers to a bundle of light rays that fail to be brought to a point or be focused at a common point. This common visual anomaly normally occurs when the cornea is not fully spherical, but actually has two radii of curvature, like a breadbasket. Accordingly, light focused by the section of the cornea having the greater curvature may be focused short of the retina, while the section with flatter curvature may provide a focal distance beyond the retina. A cylinder with focal power at the proper meridian, usually is prescribed to cancel these effects and bring light rays from various orientations to a single focus.

Anisometropia and Aniseikonia

Although refractive errors in the two eyes usually are highly correlated and very close in magnitude, when differences occur they signal a condition of unequal refraction called *anisometropia*. Each eye then requires a separate refractive-error correction. It is interesting that certain individuals can tolerate one near-sighted and one far-sighted eye, each used at the appropriate distance while the blurred imagery of the opposite eye is either tolerated, ignored, or suppressed. When each eye is purposely used in this alternating fashion, instead of utilizing bifocals, the visual status is described as a case of *monovision*.

The challenge in treating anisometropia arises from the association between the refractive correction and the change in image size it produces. Duke-Elder and Abrams (1970) estimated that a $0.25D$ difference in refractions produces about a 0.5 percent difference in the retinal image sizes of the two eyes, with a difference of 5 percent being the tolerable limit. Thus a difference in corrections in excess of about $2.0D$ produce a condition of unequal image sizes, or *aniseikonia*, that requires careful attention.

A remarkable example of the extensive adaptability of oculomotor systems (see Chapter 3) is found by examining the saccadic response to aniseikonia. Normally conjugacy, the equal movement of both eyes, is the rule. However, when each eye has a differently sized image, this would cause one eye (fovea) to land on a noncorresponding part of its image, and potentially lead to the experience of double images. Yet within a few minutes of exposure to aniseikonic images, even with a 10 percent magnification differential, disconjugacy occurs, the saccade for each eye being tailored to its particular image (Kapoula, Eggert, & Bucci, 1995).

Amblyopia and Strabismus

The term *amblyopia* literally means "blunted eye." It is thought to be a condition of selective suppression of visual function. For example, the amblyopic eye may show poor foveal acuity or the amblyope may lack stereoscopic depth perception, but color perception may be adequate.

Aside from the suppression of function, the amblyopic visual system need not exhibit any other pathologic state, but it frequently is associated with an ocular dysfunction, such as *strabismus*. This condition is sometimes referred to as a *squint* because the Greek root *strabos* referred both to looking obliquely and looking through half-closed eyelids, the combination of which is the definition of squint. Common usage reserves squint to viewing through half-closed eyelids, although not necessarily obliquely.

The strabismus anomaly, in which the eyes point in different directions, is represented in Fig. A.05. As illustrated, this produces a conflicted or rivalrous visual experience with double images of the visual field. Under these conditions, suppression of one set of images proves quite functional, because the visual conflict is thereby eliminated. This is the type of amblyopia, strabismic amblyopia, that if found early enough in childhood can be treated by patching the dominant eye. It is also the case, however, that amblyopia could arise out of an uncorrected ametropia (Duke-Elder & Wybar, 1973), and if this condition should produce a defect in the fusional motor system, then strabismus would thereby be encouraged by the amblyopia.

There is now a great deal of evidence to link strabismic amblyopia with several types of perceptual dysfunction including reduced stereo acuity, errors in eye–hand coordination (Fronius & Sireteanu, 1994), reduced ability to adapt saccade amplitudes (Bucci, Kapoula, Eggert, & Garraud, 1997), decreased interocular transfer of the motion aftereffect (O'Shea et al., 1994; Hess, Demanins, & Bex, 1997), reduced duration of motion aftereffect (Hess et al., 1997), and a large variety of monocular spatial distortions (Sireteanu, Lagreze, & Constantinescu, 1993), including inappropriate bisection and alignment of vertical and horizontal extents (Fronius & Sireteanu, 1989). It is apparent that the sequela of inadequate binocularity are numerous and significant.

Glaucoma

Glaucoma, usually associated with individuals beyond age 40, essentially represents the loss of retinal ganglion cells beginning with those that serve the retinal periphery and progressing, typically without pain, inward toward the

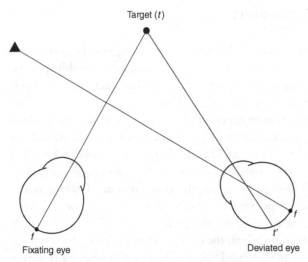

Figure A.05. *Strabismic Amblyopia with a Nasalward Deviation or Esotropia.* The circular fixation target represented on the fovea of the left eye would be imaged toward the left of the fovea in the deviated eye. According to the rules of normal retinal–cortical projection, the two images of the same object would, therefore, be seen at two different places in space. Furthermore, because the triangular object stimulates the fovea *f*, of the deviated eye, it would be seen at the same place as the circular target *t*, that stimulates the fovea of the fixating eye. There are two physiologic adaptations made in response to these potentially debilitating states. One is by way of suppression of awareness of both the target image *t*, and nontarget image in the deviated eye. Accordingly, suppression *scotomas* develop at both the fovea and periphery of the amblyopic eye. Sometimes a second solution also occurs in which the two target images come to be seen as single superimposed in the same place in space. This process, *anomalous retinal correspondence* (von Noorden, 1980, pp. 251–267), implies a shift in the localization of images of one eye in relation to that of the images in the other eye. In this example, the two circular target images would come to elicit identical visual directions with each eye, even though stimulating the fovea of the fixating eye and the periphery of the deviating eye. Likewise, under anomalous correspondence, stimulating the two foveas would lead to the experience of two different directions in space for the triangle and circle, respectively. Both adaptations are functional in that they lead to singleness of vision.

foveal ganglion cells with time. There are three diagnostic signs and symptoms associated with glaucoma, the most widely understood of which is a high intraocular pressure, greater than 22 mm mercury. The second is visual field loss leading to tunnel vision, and the third is a cupping of the optic disk as seen under slit-lamp examination. The primary mechanism for cell necrosis is, presumably, the high intraocular pressure that ultimately destroys the neurons, but it is not clear whether the pressure acts directly on the ganglion cells or indirectly on the blood vessels that supply them with nutrients.

A significant problem in identifying individuals who are at risk for glaucoma concerns individual differences in intraocular pressure, because it is known that individuals with high normative pressure may not express additional signs, whereas those with quite low pressure levels, in a condition termed *low tension glaucoma*, may exhibit cupping and visual field loss. A record of intraocular pressure taken at an early age might be helpful in establishing each individual's own base-line pressure level.

A blockage of the outflow of aqueous humor from the canal of Schlemm (see Fig. 3.03) attributable either to a displacement of the iris, in closed angle glaucoma, or by more subtle means in the open angle variety, is the basic underlying malfunction.

Heterophoria

Heterophoria is one of a set of terms devised by Stevens (1887) to describe certain nonpathological physiologic states associated with the relative directions in which the eyes point. The prefix *hetero* means "different," whereas the term *phoria* derives from "phoros," which implies "a tending" in the sense of a bearing or direction. Therefore, in heterophoria, the two eyes tend toward different directions, whereas in orthophoria the eyes are straight or true, and in hyperphoria one eye tends to point in a direction above the other. The reason for the characterization of these states as tendencies, or phorias, is that they are not noticeable under binocular vision, unlike instances of strabismus. In order to detect and measure phorias, the two eyes must be dissociated or, in modern terms, the vergence system must be open-looped so that the motor imbalance may be manifest in the absence of any sensory feedback control.

Figure A.06 demonstrates one method of opening the vergence feedback loop by placing images of dissimilar form on each retina. The reason for doing so derives from the premise that dissimilar images are inadequate stimuli for the fusion reflex, the automatic movement of the eyes that normally serves to bring similar or identically shaped images to corresponding places on each retina, and to the appropriate hemicortex as well. For example, without the Maddox cylinders in place, the disparate image of the LED on the left retina would stimulate a leftward movement of that eye, so that the LED image would fall on the fovea of both eyes. Such reflexive eye movements help to guarantee single vision by ensuring that both eyes point in the same direction,

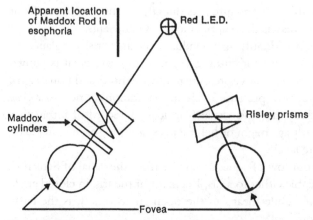

Figure A.06. *A Common Method of Measurement of Lateral Heterophoria.* (Reprinted from Ebenholtz, 1988. Sources of asthenopia in Navy flight simulators, Accession Number AD-A212699, Defense Logistics Agency, Defense Technical Information Center, Alexandria, VA.) The Maddox cylinders focus the small light-emitting diode (LED) as a line on the retina of the left eye. The Risley prisms over the left eye permit the line-image to be shifted to the fovea so as to appear coincident with the image of the LED seen with the right eye. The amount of prism shift is equivalent to the heterophoria magnitude. The technique rests on the premise that the eyes remain dissociated, in that the line is an insufficient stimulus to trigger a fusional motor response, thereby shifting the line image to the left-eye fovea. The figure represents a case of esophoria or inward turning.

at the same object, and that each half of the brain is stimulated at appropriate cortical locations. Phorias become manifest when this reflex is suspended.

Phorias are normal physiologic states, but excessive magnitudes are indicative of inordinate stress on the fusional system, which in turn may give rise to a number of undesirable symptoms collectively known as *asthenopia*.

Asthenopia

The term *asthenopia*, suggested by the Scott surgeon-oculist Mackenzie (1843), literally means "eye without strength," but it actually stands for a set of symptoms usually associated with sustained near work over prolonged time periods. Asthenopia is commonly referred to as eye strain and in contemporary times frequently has been associated with video display terminal (VDT) use. Asthenopic pain may be located in the orbits or be more diffuse as a general headache, and may implicate the neck as well as the eyebrows. Additional symptoms may include the sensation of heavy eyelids and drowsiness, hyperaesthesia of the scalp, vertigo, and gastric disturbances, including indigestion, dyspepsia, nausea, and vomiting (Duke-Elder & Abrams, 1970, pp. 566–572).

The causes of this disorder remain a source of speculation and research (Ebenholtz, 1988), but certain clinical insights about its etiology have survived over time. In essence, it is thought that oculomotor conflict and instability underlie asthenopia (Lancaster, 1932, 1943), but that purely visual symptoms such as diplopia (double vision) or obvious blur are themselves not reliable indicators of the onset of eyestrain, the reason for this being that "the condition is caused essentially by the effort to compensate for optical and muscular imperfections and if such compensation is impossible, no sustained effort is attempted" (Duke-Elder & Abrams, 1970, p. 566). Accordingly, it is the *small* errors that the oculomotor systems are capable of correcting that lie at the source of asthenopia. It is only when these systems fail that large-scale errors in control are manifest in consciousness in the form of blur, double images, and even illusions of movement. It is worth emphasizing that "it is not the error itself which causes the trouble so much as the continuous effort called forth automatically in the attempt to correct it" (Duke-Elder & Abrams, 1970, p. 564). The presence of large numbers of automatic feedback-controlled oculomotor systems, described in Chapter 3, lends great plausibility to this suggestion. It also is possible that eye muscle traction may be playing a role here that is the same as that proposed for motion sickness in Chapter 5.

Finally, it may be noted in passing that the early medical views of asthenopia may have been responsible for a fanciful tale, passed on for generations, concerning the relationship between masturbation and various pathologies, including blindness. Writing in the *Edinburgh Journal of Medicine and Surgery*, Dr. Mackenzie (1843) noted that "Palsy and insanity are not infrequent consequences of masturbation" (p. 87), and furthermore, that "I have often ascertained that asthenopia, in young men, is a result of excessive venereal indulgence, but more frequently still, of masturbation, or of involuntary emissions. I have no doubt that masturbation is a frequent cause of the same complaint in females" (p. 87). One may judge that it is through the application of scientific methods of data collection and hypothesis testing that many such spurious relationships have been eliminated from medicine.

References

Abel, L. A., Levin, S., & Holzman, P. S. (1992). Abnormalities of smooth pursuit and saccadic control in schizophrenia and affective disorders. *Vision Research, 32*, 1009–1014.

Ackrill, J. L. (Ed.). (1987). *A new Aristotle reader* (p. 214). Princeton: Princeton University Press.

Adler, F. H. (1965). *Physiology of the eye* (4th ed.). St. Louis: C. V. Mosby Co.

Allison, R. S., & Howard, I. P. (2000). Temporal dependencies in resolving monocular and binocular cue conflict in slant perception. *Vision Research, 40*, 1869–1886.

Alpern, M. (1969). Movement of the eyes. In H. Davson (Ed.), Vol. 3, *The Eye*. New York: Academic Press.

Aslin, R. N., & Dumais, S. T. (1980). Binocular vision in infants: A review and a theoretical framework. In H. W. Reese & L. P. Lipsitt (Eds.), *Advances in child development and behavior, 15*, 53–94. New York: Academic Press.

Backus, B. T., Banks, M. S., van Ee, R., & Crowell, J. A. (1999). Horizontal and vertical disparity, eye position, and stereoscopic slant perception. *Vision Research, 39*, 1143–1170.

Bacon, F. (1620). The great instauration: The distribution of the work. From B. Montague (Ed. and Trans.), *The Works*, 1854 (Vol. 3, p. 341). Philadelphia: Parry & MacMillan.

Bahill, A. T., & McDonald, J. D. (1983). Smooth pursuit eye movements in response to predictable target motions. *Vision Research, 23*, 1573–1583.

Bahill, A. T., & Stark, L. (1979). The trajectories of saccadic eye movements. *Scientific American, 240*, 108–117.

Bahill, A. T., Clark, M. R., & Stark, L. (1975). The main sequence, a tool for studying human eye movements. *Mathematical Biosciences, 24*, 191–204.

Balliet, R., & Nakayama, K. (1978). Training of voluntary torsion. *Investigative Ophthalmology and Visual Science, 17*, 303–314.

Baloh, R. W., Sharma, S., Moskowitz, H., & Griffith, R. (1979). Effect of alcohol and marijuana on eye movements. *Aviation, Space, and Environmental Medicine, 50*, 18–23.

Barnes, G. R., & Hill, T. (1984). The influence of display characteristics on active pursuit and passively induced eye movements. *Experimental Brain Research, 56*, 438–447.

Barr, C. C., Schultheis, L. W., & Robinson, D. A. (1976). Voluntary, non-visual control of the human vestibulo-ocular reflex. *Acta Otolaryngology, 81*, 365–375.

Becker, W. (1976). Do correction saccades depend exclusively on retinal feedback? A note on the possible role of non-retinal feedback. *Vision Research, 16*, 425–427.

Becker, W. (1991). Saccades. In R.H.S. Carpenter (Ed.), *Eye movements, Vision, and Visual Dysfunction.* (Vol. 1, p. 8). Boston: CRC Press.

Bedel, H. E., Klopfenstein, J. F., & Yuan, N. (1989). Extraretinal information about eye position during involuntary eye movement: Optokinetic afternystagmus. *Perception and Psychophysics, 46*, 579–586.

Bell, C. (1811). Idea of a new anatomy of the brain: Submitted for the observation of his friends. (privately printed), London. Reprinted in R. J. Herrnstein, & E. G. Boring, (Eds.). *A sourcebook in the history of psychology*, 1965, pp. 17–19. Cambridge, MA: Harvard University Press.

Bell, C. (1826). On the nervous circle which connects the voluntary muscles with the brain. *Philosophical Transactions of the Royal Society*, 163–173. Cited in E. G. Boring, (1942), *Sensation and perception in the history of experimental psychology.* pp. 526–527, p. 565. New York: Appleton-Century-Crofts.

Berkeley, G. (1709/1948). An essay towards a new theory of vision. In A. A. Luce & T. E. Jessop (Eds.), *The works of George Berkeley Bishop of Cloyne, Vol. I.*, New York: Thomas Nelson & Sons Ltd.

Bernard, C. (1865/1957). *An introduction to the study of experimental medicine.* (H. C. Greene, Trans.). New York: Dover Publications, Inc.

Bielschowsky, A. (1938). Lectures on motor anomalies. *American Journal of Ophthalmology, 21*, 843–854.

Bles, W. (1979). *Sensory interactions and human posture: An experimental study.* Unpublished doctoral dissertation, Vrije Universiteit te Amsterdam.

Bles, W. (1981). Stepping around: Circular vection and coriolis effects. In J. Long & A. Baddeley (Eds.), *Attention and Performance IX.* Hillsdale, NJ: Lawrence Erlbaum Associates.

Blinkov, S. M., & Glezer, I. I. (1968). *The human brain in figures and tables: A quantitative handbook* (p. 120). B. Haugh (Trans.). New York: Plenum Press & Basic Books, Inc.

Bloomberg, J. J., Merkle, L. A., Barry, S. R., Huebner, W. P., Cohen, H. S., Mueller, S. A., & Fordice, J. (2000). Effects of adaptation of vestibulo-ocular reflex function on manual target localization. *Journal of Vestibular Research, 10*, 75–86.

Boring, E. G. (1942). *Sensation and perception in the history of experimental psychology.* New York: Appleton-Century Crofts.

Boring, E. G. (1950). *A history of experimental psychology.* New York: Appleton-Century-Crofts, Inc.

Bouwhuis, D. G., Bridgeman, B., Owens, D. A. Shebilske, W. L., & Wolff, P. (Eds.) (1987). *Sensorimotor interactions in space perception and action.* New York: Elsevier Science.

Bradley, D. C., Maxwell, M., Andersen, R. A., Banks, M. S., & Shenoy, K. V. (1996). Mechanisms of heading perception in primate visual cortex. *Science, 273*, 1544–1547.

Bradshaw, M. F., & Rogers, B. J. (1996). The interaction of binocular disparity and motion parallax in the computation of depth. *Vision Research, 36*, 3457–3468.

Brandt, T., Wist, E. R., & Dichgans, J. (1975). Foreground and background in dynamic spatial orientation. *Perception and Psychophysics, 17*, 497–503.

Braunstein, M. L. (1976). *Depth perception through motion.* New York: Academic Press.

Bridgeman, B. (1996). Extraretinal signals in visual orientation. In W. Prinz, & B. Bridgeman (Eds.), *Handbook of perception and action* (Vol. 1, pp. 191–222). New York: Academic Press.

Bridgeman, B., & Graziano, J. A. (1989). Effect of context and efference copy on visual straight-ahead. *Vision Research, 29*, 1729–1736.

Bridgeman, B., & Stark, L. (1981). Efferent copy and visual direction. *Investigative Ophthalmology and Visual Science (Suppl.) 20*, 55.

Brindley, G. S., & Lewin, W. S. (1968). The sensation produced by electrical stimulation of the visual cortex. *Journal of Physiology* (London), *196*, 479–493.

Brown, J. L. (1965). After images. In C. H. Graham, (Ed.), *Vision and Visual Perception* (Chap. 17), New York: John Wiley & Sons.

Bruell, J. H., & Albee, G. W. (1955). Notes toward a motor theory of visual egocentric localization. *Psychological Review, 62*, 391–400.

Bucci, M. P., Kapoula, Z., Eggert, T., & Garraud, L. (1997). Deficiency of adaptive control of the binocular coordination of saccades in strabismus. *Vision Research, 37*, 2767–2777.

Bucher, U. J., Mast, F., & Bischof, N. (1992). An analysis of ocular counterrolling in response to body positions in the three-dimensional space. *Vestibular Research, 2,* 213–220.

Butterworth, G., & Henty, C. (1991). Origins of the proprioceptive function of vision: Visual control of posture in one day old domestic chicks. *Perception, 20,* 381–386.

Callan, J. W., & Ebenholtz, S. M. (1982). Directional changes in the vestibular ocular response as a result of adaptation to optical tilt. *Vision Research, 22,* 37–42.

Carpenter, R. H. S. (1977). *Movements of the Eyes.* London: Pion Limited.

Carter, D. B. (1963). Effects of prolonged wearing of prism. *American Journal of Optometry, 40,* 265–273.

Chalmers, D. J. (1995). Facing up to the problem of consciousness. *Journal of Consciousness Studies, 2* (3), 200–219.

Cheung, B. S. K., Howard, I. P., & Money, K. E. (1991). Visually-induced sickness in normal and bilaterally labyrinthine-defective subjects. *Aviation, Space, and Environmental Medicine, 62,* 527–531.

Cheung, B. S. K., Howard, I. P., Nedzelski, J. M., & Landolt, J. P. (1989). Circularvection about earth-horizontal axes in bilateral labyrinthine-defective subjects. *Acta Otolaryngology* (Stockholm), *108,* 336–344.

Citek, K., & Ebenholtz, S. M. (1996). Vertical and horizontal eye displacement during static pitch and roll postures. *Vestibular Research, 6,* 213–228.

Cohen, M. M., Crosbie, R. J., & Blackburn, L. H. (1973). Disorienting effects of aircraft catapult launchings. *Aerospace Medicine, 44,* 37–39.

Cohen, M. M., Ebenholtz, S. M., & Linder, B. J. (1995). Effects of optical pitch on oculomotor control and the perception of target elevation. *Perception and Psychophysics, 57,* 433–440.

Collewijn, H., Martins, A. J., & Steinman, R.M. (1983). Compensatory eye movements during active and passive head movements. Fast adaptation to changes in visual magnification. *Journal of Physiology,* (London), 340, 259–286.

Collins, W. E. (1966). Vestibular responses from figure skaters. *Aerospace Medicine, 37,* 1098–1104.

Cooper, J., Selenow, A., Ciuffreda, K. J., Feldman, J., Faverty, J., Hokoda, S. C., & Silver, J. (1983). Reduction of asthenopia in patients with convergence insufficiency after fusional vergence training. *American Journal of Optometry and Physiological Optics, 60,* 982–989.

Cooper, S., & Daniel, P. M. (1949). Muscle spindles in human extrinsic muscles. *Brain, 72,* 1–24.

Cooper, S., Daniel, P. M., & Whitteridge, D. (1955). Muscle spindles and other sensory endings in the extrinsic eye muscles: The physiology and anatomy of these receptors and their connections with the brain stem. *Brain, 78,* 564–583.

Coren, S. (1969). Brightness contrast as a function of figure-ground relations. *Journal of Experimental Psychology, 80,* 517–524.

Coren, S., Bradley, D. R., Hoenig, P., & Girgus, J. S. (1975). The effect of smooth tracking and saccadic eye movements on the perception of size: The shrinking circle illusion. *Vision Research, 15,* 49–55.

Cormack, R. H. (1984). Stereoscopic depth perception at far viewing distances. *Perception and Psychophysics, 35,* 423–428.

Cornsweet, T. N., & Crane H. D. (1973). Research note: Training the visual accommodation system. *Vision Research, 13,* 713–715.

Craske, B. (1977). Perception of impossible limb positions induced by tendon vibration. *Science, 196,* 71–73.

Crone, R. A. (1975). Optically induced eye torsion II. Optostatic and optokinetic cycloversion. *Albrecht von Graefes Archiv fur Klinische und Experimentelle Ophthalmologie, 196,* 1–7.

Curtin, B. J. (1985). *The myopias* (pp. 61–151). New York: Harper & Row.

Cutting, J. E. (1986). *Perception with an eye for motion.* Cambridge, MA: MIT Press.

Dai, M., McGarvie, L., Kozlovskaya, I., Raphan, T., & Cohen, B. (1994). Effects of spaceflight on ocular counterrolling and the spatial orientation of the vestibular system. *Experimental Brain Research, 102,* 45–56.

Daunton, N., & Thomsen, D. (1979). Visual modulation of otolith units in cat vestibular nuclei. *Experimental Brain Research, 37,* 173–176.

Davson, H. (1969). The intraocular fluids. In H. Davson (Ed.), *The eye, Vol.* 1. (2nd ed.), New York: Academic Press.

de Coriolis, G. G. (1835). *Memoire sur les equations du movement relatif des systems de corp.* Paris.

Delorme, A., Frigon, J., & Lagace, C. (1989). Infants' reactions to visual movement of the environment. *Perception, 18,* 667–673.

Descartes, R. (1637/1965). *Discourse on method, optics, geometry, and meteorology.* (P. J. Olscamp, Trans.). New York: Bobbs-Merrill.

Descartes, R. (1664/1972). *Treatise of man* (Thomas Steele Hall, Trans.). Cambridge: MA, Harvard University Press. Original work. L'Homme de Rene Descartes (1664). Paris: Charles Angot.

Diamond, S. G., & Markham, C. H. (1983). Ocular counterrolling as an indicator of vestibular otolith function. *Neurology, 33,* 1460–1469.

Diamond, S. G., & Markham, C. H. (1991). Prediction of space motion sickness susceptibility by disconjugate eye torsion in parabolic flight. *Aviation, Space, and Environmental Medicine, 62,* 201–205.

Dichgans, J., & Brandt, T. (1978). Visual-vestibular interaction: effects on self-motion perception and postural control. In R. Held, H. W. Leibowitz, and H.-L. Teuber (Eds.), *Handbook of sensory physiology: Perception* (Vol. 8, pp. 755–804). New York: Springer-Verlag.

Dichgans, J., Held, R., Young, L. R., & Brandt, T. (1972). Moving visual scenes influence the apparent direction of gravity. *Science, 178*, 1217–1219.

Dix, M. R., & Hood, J. D. (1969). Observations upon the nervous mechanism of vestibular habituation. *Acta Otolaryngologica, 67*, 310–318.

Dubois, P. H. (1982). *Effect of vestibulo-ocular retraining on object position constancy: prediction by vector analysis.* Unpublished master's thesis, Department of Psychology, University of Wisconsin, Madison, WI.

Duke-Elder, Sir S., & Abrams, D. (1970). *System of ophthalmology.* Sir S. Duke-Elder (Ed.), *Ophthalmic optics and refraction.* Vol. 5. St. Louis: C. V. Mosby. Co.

Duke-Elder, Sir S., & Wybar, K. (1973). *System of ophthalmology.* Sir S. Duke-Elder (Ed.), *Ocular motility and strabismus*, (Vol. 6). St. Louis: C.V. Mosby Co.

Duncker, K. (1929). Über induzierte bewegung. In W. D. Ellis (Ed. and Trans.) *A sourcebook of Gestalt Psychology* (1955). New York: The Humanities Press Inc.

Ebenholtz, S. M. (1970). Perception of the vertical with body tilt in the median plane. *Journal of Experimental Psychology, 83*, 1–6.

Ebenholtz, S. M. (1976). Additivity of aftereffects of maintained head and eye rotations: An alternative to recalibration. *Perception and Psychophysics, 19*, 113–116.

Ebenholtz, S. M. (1977a). The constancies in object orientation: An algorithm processing approach. In W. Epstein (Ed)., *Stability and constancy in visual perception: Mechanisms and process.* New York: John Wiley & Sons.

Ebenholtz, S. M. (1977b). Determinants of the rod and frame effect: The role of retinal size. *Perception and Psychophysics, 22*, 531–538.

Ebenholtz, S. M. (1978). Aftereffects of sustained vertical divergence: Induced vertical phoria and illusory target height. *Perception, 7*, 305–314.

Ebenholtz, S. M. (1981). Hysteresis effects in the vergence control system: perceptual implications. In D. F. Fisher, R. A. Monty, & J. W. Senders (Eds.), *Eye movements: Cognition and visual perception*, Hillsdale, NJ: Lawrence Erlbaum Associates.

Ebenholtz, S. M. (1983). Accomodative hysteresis: A precursor for induced myopia? *Investigative Ophthalmology and Visual Science, 24*, 513–515.

Ebenholtz, S. M. (1984a). Perceptual coding and adaptations of the oculomotor systems. In L. Spillman & B. R. Wooten, (Eds.), *Sensory experience, adaptation, and perception: Festschrift for Ivo Kohler* (Chap. 19). Hillsdale, NJ: Lawrence Erlbaum Associates, Inc.

Ebenholtz, S. M. (1984b). Oculomotor adaptive systems. In M. E. McCauley (Ed.), *Research issues in simulator sickness: Proceedings of a workshop* (pp. 33–41). *Committee on Human Factors*, National Research Council, Washington, DC: National Academy Press.

Ebenholtz, S. M. (1985). Accommodative hysteresis: Relation to resting focus. *American Journal of Optometry and Physiological Optics, 62,* 755–762.

Ebenholtz, S. M. (1986). Properties of adaptive oculomotor control systems and perception. *Acta Psychologica, 63,* 233–246.

Ebenholtz, S. M. (1988). *Sources of asthenopia in Navy flight simulators.* (Accession Number AD-A212699.) Defense Logistics Agency, Defense Technical Information Center, Alexandria, VA.

Ebenholtz, S. M. (1990). Metamorphosis from rod and frame to visual-vestibular interaction. In I. Rock (Ed.), *The legacy of Solomon Asch: Essays in cognition and social psychology,* Hillsdale, NJ: Lawrence Erlbaum Associates.

Ebenholtz, S. M. (1991). The effects of teleoperator-system displays on human oculomotor systems. *Society of Automotive Engineers, Inc., Technical Paper Series 911391,* 1–21.

Ebenholtz, S. M. (1992). Accommodative hysteresis as a function of target-dark focus separation. *Vision Research, 32,* 925–929.

Ebenholtz, S. M., & Citek, K. (1992). Absence of adaptive plasticity after voluntary vergence and accommodation. *Investigative Ophthalmology and Visual Science* (Abstract), *33,* 1101.

Ebenholtz, S. M., & Citek, K. (1995). Absence of adaptive plasticity after voluntary vergence and accommodation. *Vision Research, 35,* 2773–2783.

Ebenholtz, S. M., Cohen, M. M., & Linder, B. J. (1994). The possible role of nystagmus in motion sickness: A hypothesis. *Aviation, Space, and Environmental Medicine, 65,* 1032–1035.

Ebenholtz, S. M., & Fisher, S. K. (1982). Distance adaptation depends upon plasticity in the oculomotor control system. *Perception and Psychophysics, 31,* 551–560.

Ebenholtz, S. M., & Paap, K. R. (1973). The constancy of object orientation: Compensation for ocular rotation. *Perception and Psychophysics, 14,* 458–470.

Ebenholtz, S. M., & Paap, K. R. (1976). Further evidence for an orientation constancy based upon registration of ocular position. *Psychological Research, 38,* 395–409.

Ebenholtz, S. M., & Shebilske, W. (1975). The doll reflex: Ocular counterrolling with head-body tilt in the median plane. *Vision Research, 15,* 713–717.

Ebenholtz, S. M., & Wolfson, D. M. (1975). Perceptual aftereffects of sustained convergence. *Perception and Psychophysics, 17,* 485–491.

Ellerbrock, V. J., & Fry, G. A. (1941). The aftereffect induced by vertical divergence. *American Journal of Optometry, 18,* 450–454.

Emmert, E. (1881). Grössenverhältnisse der Nachbilder. *Klinische Monatsblätter Augenheilkunde, 19,* 443–450.

Enright, J. T. (1984). Changes in vergence mediated by saccades. *Journal of Physiology* (London), *350,* 9–31.

Enright, J. T. (1992). The remarkable saccades of asymmetrical vergence. *Vision Research, 322,* 2261–2276.

Enright, J. T. (1996). Slow-velocity assymetrical convergence: A decisive failure of "Hering's Law." *Vision Research, 36,* 3667–3684.

Erkelens, C. J. (1987). Adaptation of ocular vergence to stimulation with large disparities. *Experimental Brain Research, 66,* 507–516.

Erkelens, C. J. (2000). Perceived direction during monocular viewing is based on signals of the viewing eye only. *Vision Research, 40,* 2411–2419.

Evanoff, J. N., & Lackner, J. R. (1987). Influence of maintained ocular deviation on the spatial displacement component of the oculogyral illusion. *Perception and Psychophysics, 42,* 25–28.

Festinger, L., Ono, H., Burnham, C. A., & Bamber, D. (1967). Efference and the conscious experience of perception. *Journal of Experimental Psychology (Monograph), 74* (4, Whole No. 637), 1–36.

Festinger, L., Sedgwick, H. A., & Holtzman, J. D. (1976). Visual perception during smooth pursuit eye movements. *Vision Research, 16,* 1377–1386.

Fincham, E. F. (1962). Accommodation and convergence in the absence of retinal images. *Vision Research, 1,* 425–440.

Fisher, K., & Sanchez, N. (1997). The effect of accommodative hysteresis on apparent stationarity. *Ophthalmic and Physiological Optics, 17,* 112–121.

Fisher, R. F. (1986). The ciliary body in accommodation. *Transactions of the Ophthalmology Society.* U.K., *105,* 208–219.

Fisher, S. K., & Ciuffreda, K. J. (1988). Accommodation and apparent distance. *Perception, 17,* 609–621.

Fisher, S. K., & Ciuffreda, K. J. (1989). The effect of accommodative hysteresis on apparent distance. *Ophthalmic and Physiological Optics, 9,* 184–190.

Fisher, S. K., & Ciuffreda, K. J. (1990). Adaptation to optically increased interocular separation under naturalistic viewing conditions. *Perception, 19,* 171–180.

Fisher, S. K., Ciuffreda, K. J., Tannen, B., & Super, P. (1988). Stability of tonic vergence. *Investigative Ophthalmology and Visual Science, 29,* 1577–1581.

Fisher, S. K., & Ebenholtz, S. M. (1986). Does perceptual adaptation to telestereoscopically enhanced depth depend on the recalibration of binocular disparity? *Perception and Psychophysics, 40*, 101–109.

Fischer, M. H., & Kornmüller, A. E. (1930). Optokinetisch ausgeloste bewegungswahrnehmungen und optokinetischer nystagmus. *Journal fur Psychologie und Neurologie, 41*, 273–308.

Fleischl, E. (1882). Physiologisch-optische Notizen. *Sitzungsberichte der Akademie der Wissenschaften, 86*, 17–25.

Flom, M. C. (1960). On the relationship between accommodation and accommodative convergence (Part 1). *American Journal of Optometry, 37*, 474–482.

Foley, J. M. (1980). Binocular distance perception. *Psychological Review, 87*, 411–434.

Fried, A. H. (1973). Convergence as a cue to distance. (Doctoral dissertation, New School for Social Research.) Cited in W. Epstein (Ed.), *Stability and constancy in visual perception* (1977), H. Ono, & J. Cummerford, *Stereoscopic depth constancy*, (Chap. 4), pp. 107–108. New York: Wiley.

Fronius, M., & Sireteanu, R. (1989). Monocular geometry is selectively distorted in the central visual field of strabismic amblyopes. *Investigative Ophthalmology and Visual Science, 30*, 2034–2044.

Fronius, M., & Sireteanu, R. (1994). Pointing errors in Strabismus: Complex patterns of distorted visuomotor coordination. *Vision Research, 34*, 689–707.

Gauthier, G. M., Nommay, D., & Vercher, J.-L. (1990). The role of ocular muscle proprioception in visual localization of targets. *Science, 249*, 58–61.

Gauthier, G. M., & Robinson, D. A. (1975). Adaptation of humans' vestibulo-ocular reflex to magnifying lenses. *Brain Research, 92*, 331–335.

Gibson, J. J. (1950). *The perception of the visual world*. Cambridge: The Riverside Press.

Gibson, J. J. (1966). *The senses considered as perceptual systems*. Boston: Houghton Mifflin.

Gibson, J. J. (1979). *The ecological approach to visual perception*. Boston: Houghton Mifflin.

Gilchrist, A. L. (1977). Perceived lightness depends on perceived spatial arrangement. *Science, 95*, 185–187.

Glasser, A., & Campbell, M. C. W. (1998). Presbyopia and the optical changes in the human crystalline lens with age. *Vision Research, 38*, 209–229.

Gnadt, J. W. (1992). The relationship of accommodation and vergence during voluntary near response in the dark in monkeys. *Investigative Ophthalmology and Visual Science* (Abstract), *33*, 1100.

Gogel, W. C. (1973). The organization of perceived space. I. Perceptual interactions. *Psychologische Forschung, 36*, 115–221.

Gogel, W. C., & Tietz, J. D. (1974). The effect of perceived distance on perceived movement. *Perception and Psychophysics, 16*, 70–78.

Gonshor, A., & Melvill Jones, G. (1976). Extreme vestibulo-ocular adaptation induced by prolonged optical reversal of vision. *Journal of Physiology, 256*, 381–414.

Goodwin, G. M., McCloskey, D. I., & Mathews, P. B. C. (1972a). The contribution of muscle afferents to kinesthesia shown by vibration induced illusions of movement and by the effects of paralyzing joint afferents. *Brain, 95*, 705–748.

Goodwin, G. M., McCloskey, D. I., & Mathews, P. B. C. (1972b). Proprioceptive illusions induced by muscle vibration: Contribution by muscle spindles to perception? *Science, 175*, 1382–1384.

Gordon, C. R., Spitzer, O., Doweck, I., Shupak, A., & Gadoth, N. (1996). The vestibulo-ocular reflex and seasickness susceptibility. *Journal of Vestibular Research, 6*, 229–233.

Granit, R. (1970). *The basis of motor control* (Chap. 7). New York: Academic Press.

Graybiel, A., & Hupp, D. I. (1946). The oculo-gyral illusion, a form of apparent motion which may be observed following stimulation of the semicircular canals. *Journal of Aviation Medicine, 17*, 3–27.

Graybiel, A., & Kellogg, R. S. (1967). The inversion illusion in parabolic flight: Its probable dependence on otolith function. *Aerospace Medicine, 38*, 1099–1103.

Graybiel, A., & Knepton, J. (1976). Sopite syndrome: A sometimes sole manifestation of motion sickness. *Aviation, Space, and Environmental Medicine, 47*, 873–882.

Grossman, G. E., & Leigh, R. J. (1990). Instability of gaze during locomotion in patients with deficient vestibular function. *Annals of Neurology, 27*, 528–532.

Grossman, G. E., Leigh, R. J., Abel, L. A., Lanska, D. J., & Thurston, S. E. (1988). Frequency and velocity of rotational head perturbations during locomotion. *Experimental Brain Research, 70*, 470–476.

Grossman, G. E., Leigh, R. J., Bruce, E. N., Huebner, W. P., & Lanska, D. J. (1989). Performance of the human vestibulo-ocular reflex during locomotion. *Journal of Neurophysiology, 62*, 264–272.

Grüsser, O.-J., Krizic, A., & Weiss, L.-R. (1987). Afterimage movement during saccades in the dark. *Vision Research, 27*, 215–226.

Gullstrand, A. (1924/1962). The optical system of the eye. In J. P. C. Southall (Ed. and Trans.), *Helmholtz's treatise on physiological optics* (Vol. 1, pp. 350–358). From the 3rd German Edition. New York: Dover Publications.

Gwiazda, J., Thorn, F., Bauer, J., & Held, R. (1991). Tonic accommodation increases more in myopic than hyperopic children following near work. *Investigative Ophthalmology and Visual Science* (Abstract), *32*, 1125.

Hagbarth, K. E., & Eklund, G. (1966). Motor effects of vibratory stimuli in man. In R. Granit (Ed.), *Muscular afferents and motor control* (pp. 177–186), Stockholm: Almquist and Wiksell.

Hallett, P. E., & Lightstone, A. D. (1976). Saccadic eye movements to flashed targets. *Vision Research, 16*, 107–114.

Hartmann, W. M. (1999). How we localize sound. *Physics Today, 52*, 24–29.

Heckmann, T., & Howard, I. P. (1991). Induced motion: Isolation and dissociation of egocentric and vection-entrained components. *Perception, 20*, 285–305.

Heckmann, T., & Post, R. B. (1988). Induced motion and optokinetic after nystagmus: Parallel response dynamics with prolonged stimulation. *Vision Research, 28*, 681–694.

Hein, A., & Diamond, R. (1983). Contribution of eye movement to the representation of space (Chap. 7). In A. Hein & M. Jeannerod (Eds.), *Spatially oriented behavior*. New York: Springer-Verlag.

Held, R. (1968). Dissociation of visual functions by deprivation and rearrangement. *Psychologische Forschung, 31*, 338–348.

Held, R., & Hein, A. (1963). Movement-produced stimulation in the development of visually guided behavior. *Journal of Comparative and Physiological Psychology, 56*, 872–876.

Helmholtz, H. von (1910/1962). *Treatise on physiological optics*, Vol. 1, Republication of translation of the Third German Edition (J. P. C. Southall, Ed., *Optical Society of America*, 1925). New York: Dover.

Helmholtz, H. von (1910/1962). *Treatise on physiological optics*, Vol. 3, Republication of translation of the Third German Edition (J. P. C. Southall, Ed., *Optical Society of America*, 1925). New York: Dover.

Henn, V., Young, L. R., & Finley, C. (1974). Vestibular nucleus units in alert monkeys are also influenced by moving visual scenes. *Brain Research, 71*, 144–149.

Henriksson, N. G., Pyykko, I., Schalen, L., & Wennmo, C. (1980). Velocity patterns of rapid eye movements. *Acta Otolaryngology, 89*, 504–512.

Henson, D. B. (1978). Corrective saccades: Effects of altering visual feedback. *Vision Research, 18*, 63–67.

Hering, E. (1868). *Die lehre vom binocularen sehen*. Leipzig: Engleman.

Hering, E. (1879/1942). *Spatial sense and movements of the eye* (pp. 38–41). Carl A. Radde, (Trans.), Baltimore: The American Academy of Optometry.

Heron, G., Smith, A. C., & Winn, B. (1981). The influence of method on the stability of dark focus position of accommodation. *Ophthalmic and Physiological Optics, 1,* 79–90.

Hess, R. F., Demanins, R., & Bex, P. J. (1997). A reduced motion aftereffect in strabismic amblyopia. *Vision Research, 37,* 1303–1311.

Hildreth, E. C. (1992). Recovering heading for visually-guided navigation. *Vision Research, 32,* 1177–1192.

Hilgard, E. R. (1978). Pain perception in man (Chap. 27). In R. Held, H. W. Leibowitz, & H.-L. Teuber (Eds.), *Handbook of sensory physiology, Vol. VIII, Perception.* New York: Springer-Verlag.

Hofsten, C. von (1977). Binocular convergence as a determinant of reaching behavior in infancy. *Perception, 6,* 139–144.

Honda, H. (1990). The extraretinal signal from the pursuit-eye-movement system: Its role in the perceptual and the egocentric localization systems. *Perception and Psychophysics, 48,* 509–515.

Hood, J. D. (1984). Tests of vestibular function. In M. R. Dix & J. D. Hood (Eds.), *Vertigo* (Chap. 3). New York: John Wiley & Sons.

Houchin, K. W., Dunbar, J. A., & Lingua, R. W. (1992). Reduction of postoperative emesis in medial rectus muscle surgery. *Investigative Ophthalmology and Visual Science, 33,* 1336.

Howard, I. P. (1982). *Human visual orientation.* New York: John Wiley.

Howard, I. P., & Gonzalez, E. G. (1987). Human optokinetic nystagmus in response to moving binocularly disparate stimuli. *Vision Research, 27,* 1807–1816.

Howard, I. P., & Howard, A. (1994). Vection: The contributions of absolute and relative visual motion. *Perception, 23,* 745–751.

Hu, S., Davis, M. S., Klose, A. H., Zabinski, E. M., Meux, S. P., Jacobsen, H. A., Westfall, J. M., & Gruber, M. B. (1997). Effects of spatial frequency of a vertically striped rotating drum on vection-induced motion sickness. *Aviation, Space, and Environmental Medicine, 68,* 306–311.

Hu, S., McChesney, K. A., Player, K. A., Bahl, A. M., Buchanan, J. B., & Scozzafava, J. E. (1999). Systematic investigation of physiological correlates of motion sickness induced by viewing an optokinetic rotating drum. *Aviation, Space, and Environmental Medicine, 70,* 759–765.

Hudspeth, A. J. (1983). The hair cells of the inner ear. *Scientific American, 248,* 54–64.

Hudspeth, A. J., & Jacobs, R. (1979). Stereocilia mediate transduction in vertebrate hair cells. *Proceedings National Academy of Science, 76,* 1506–1509.

Hughes, A. (1972). Vergence in the cat. *Vision Research, 12,* 1961–1994.

Hughes, A. (1977). The topography of vision in mammals of contrasting life styles. In F. Crescitelli (Ed.), *Handbook of Sensory Physiology, Vol. VII/5* (pp. 614–644). Berlin: Springer.

Ilg, U. J., Bridgeman, B., & Hoffman, K. P. (1989). Influence of mechanical disturbance on oculomotor behavior. *Vision Research, 29*, 545–551.

Indiveri, G., & Douglas, R. (2000). Neuromorphic vision sensors. *Science, 288*, 1189–1190.

Ingle, D., Schneider, G., Trevarthen, C., & Held, R. (1967). Locating and identifying: Two modes of visual processing, a symposium. *Psychologische Forschung, 31*, 42–43.

Israel, I., Sievering, D., & Koenig, E. (1995). Self-rotation estimate about the vertical axis. *Acta Otolaryngologica* (Stockholm), *115*, 3–8.

Isu, N., Yanagihara, M., Mikuni, T., & Koo, J. (1994). Coriolis effects are principally caused by gyroscopic angular acceleration. *Aviation, Space, and Environmental Medicine, 65*, 627–631.

Isu, N., Yanagihara, M., Yoneda, S., Hattori, K., & Koo, J. (1996). The severity of nauseogenic effect of cross-coupled rotation is proportional to gyroscopic angular acceleration. *Aviation, Space, and Environmental Medicine, 67*, 325–332.

Ittelson, W. H., & Ames, A. Jr., (1950). Accommodation, convergence and their relation to apparent distance. *Journal of Psychology, 30*, 43–67.

J. C. (1952). Living without a balancing mechanism. *The New England Journal of Medicine, 246*, 458–460.

James, W. (1890/1950). *The principles of psychology*. New York: Dover Publications.

Jampel, R. S. (1959). Representation of the near-response on the cerebral cortex of the Macaque. *American Journal of Ophthalmology, 48*, 573–582.

Jones, M. B. (1965). Individual variations in the postocular lines of regard. *American Journal of Psychology, 78*, 627–633.

Jordan, S. (1970). Ocular pursuit movements as a function of visual and proprioceptive stimulation. *Vision Research, 10*, 775–780.

Kalil, R. E., & Freedman, S. J. (1966). Persistence of ocular rotation following compensation for displaced vision. *Perceptual and Motor Skills, 22*, 135–139.

Kapoula, Z., Eggert, T., & Bucci, M. P. (1995). Immediate saccade amplitude disconjugacy induced by unequal images. *Vision Research, 35*, 3505–3518.

Katz, R. L., & Bigger, J. T. (1970). Cardiac arrhythmias during anesthesia and operation. *Anesthesiology, 33*, 193–213.

Kaufman, L. (1974). *Sight and mind* (Chap. 1). York: Oxford.

Kaufman, L., & Kaufman, J. H. (2000). Explaining the moon illusion. *Proceedings of the National Academy of Sciences, 97*, 500–505.

Kaufman, L., & Rock, I. (1962). The moon illusion, I. *Science, 136*, 953–961.

Keating, M. P. (1988). *Geometric, physical, and visual optics.* Boston: Butterworths.

Keele, C. A., & Neil, E. (1965). *Samson Wright's applied physiology* (11th ed.). New York: Oxford University Press.

Kennedy, R. S., Graybiel, A., McDonough, R. C., & Becksmith, Fr. D. (1968). Symptomatology under storm conditions in the North Atlantic in control subjects and in persons with bilateral labyrinthine defects. *Acta Otolaryngologica, 66,* 533–540.

Kennedy, R. S., Hettinger, L. J., Harm, D. L., Ordy, M., & Dunlop, W. P. (1996). Psychophysical scaling of circular vection (CV) produced by optokinetic (OKN) motion: Individual differences and effects of practice. *Journal of Vestibular Research, 6,* 331–341.

Kerwin, J. (1977). Skylab 2 crew observations and summary. In R. S. Johnson & L. F. Dietlein (Eds.), *Biomedical Results from Skylab.* NASA SP-377, 27–29, Washington, DC: NASA.

Khater, T. T., Baker, J. F., & Peterson, B. W. (1990). Dynamics of adaptive change in human vestibular-ocular reflex direction. *Journal of Vestibular Research, 1,* 23–29.

Koffka, K. (1935). *Principles of Gestalt psychology.* New York: Harcourt, Brace & Co., Inc.

Kommerell, G., Olivier, D., & Theopold, H. (1976). Adaptive programming of phasic and tonic components in saccadic eye movements. Investigations in patients with abducens palsy. *Investigative Ophthalmology, 15,* 657–660.

Kotulak, J. C., & Schor, C. M. (1986a). The accommodative response to subthreshold blur and to perceptual fading during the Troxler phenomenon. *Perception, 15,* 7–15.

Kotulak, J. C., & Schor, C. M. (1986b). The dissociability of accommodation from vergence in the dark. *Investigative Ophthalmology and Visual Science, 27,* 544–551.

Kruger, P. B., & Pola, J. (1985). Changing target size is a stimulus for accommodation. *Journal of the Optical Society of America A, 75,* 1832–1835.

Lackner, J. R. (1975). Pursuit eye movements elicited by muscle afferent information. *Neuroscience Letters, 1,* 25–28.

Lackner, J. R. (1977). Induction of illusory self-rotation and nystagmus by a rotating sound-field. *Aviation, Space, and Environmental Medicine, 65,* 1032–1035.

Lackner, J. R. (1978). Some mechanisms underlying sensory and postural stability in man. In R. Held, H. W. Leibowitz, & H.-L. Teuber (Eds.). *Handbook of sensory physiology, Perception* (Vol. 8). New York: Springer-Verlag.

Lackner, J. R., & Graybiel, A. (1983). Perceived orientation in free-fall depends on visual, postural, and architectural factors. *Aviation, Space, and Environmental Medicine, 54,* 47–51.

Lancaster, W. B. (1932). Ocular symptoms of faulty illumination. *American Journal of Ophthalmology, 75,* 783–788.

Lancaster, W. B. (1943). The story of asthenopia: Important part played by Philadelphia. *Archives of Ophthalmology, 30,* 167–178.

Lee, D. N., & Aronson E. (1974). Visual proprioceptive control of standing in human infants. *Perception and Psychophysics, 15,* 529–532.

Leibowitz, H., & Moore, D. (1966). Role of changes in accommodation and convergence in the perception of size. *Journal of the Optical Society of America, 56,* 1120–1123.

Leibowitz, H. W., & Owens, D. A. (1978). New evidence for the intermediate position of relaxed accommodation. *Documenta Ophthalmologica, 46,* 133–147.

Leibowitz, H. W., & Post, R. B. (1982). The two modes of processing concept and some implications. In J. J. Beck (Ed.), *Organization and representation in perception.* New York: Erlbaum.

Leigh, R. J., & Zee, D. S. (1984). *The neurology of eye movements.* Philadelphia: F.A. Davis Co.

Lewald, J., & Ehrenstein, W. H. (2000). Visual and proprioceptive shifts in perceived egocentric direction induced by eye-position. *Vision Research, 40,* 539–547.

Li, W., & Matin, L. (1998). Change in visually perceived eye level without change in perceived pitch. *Perception, 27,* 553–572.

Lichtenberg, B. K., Young, L. R., & Arrott, A. P. (1982). Human ocular counterrolling induced by varying linear accelerations. *Experimental Brain Research, 48,* 127–136.

Lindberg, D. C. (1976). *Theories of vision from Al-kindi to Kepler.* Chicago: The University of Chicago Press.

Lindberg, D. C. (1978a). The transmission of Greek and Arabic learning to the West. In D. C. Lindberg (Ed.), *Science in the middle ages* (pp. 52–90). Chicago: The University of Chicago Press.

Lindberg, D. C. (1978b). The science of optics. In D. C. Lindberg (Ed.), *Science in the middle ages* (pp. 338–368). Chicago: The University of Chicago Press.

Linksz, A. (1950). *Physiology of the eye,* Vol. 1. New York: Grune & Stratton.

Linksz, A. (1952). *Physiology of the eye,* Vol. 2. New York: Grune & Stratton.

Lisberger, S. G. (1984). The latency of pathways containing the site of motor learning in the monkey vestibulo-ocular reflex. *Science, 225,* 74–76.

Livingstone, M. S., & Hubel, D. H. (1987). Psychophysical evidence for separate channels for the perception of form, color, movement, and depth. *Journal of Neuroscience, 7,* 3416–3468.

Lowenstein, O., & Loewenfeld, I. (1959). Influence of retinal adaptation upon the pupillary reflex to light in normal man. *American Journal of Ophthalmology, 48*, 536–549.

Lowenstein, O., & Loewenfeld, I. E. (1969). The pupil. In H. Davson (Ed.), *The eye*, Vol. 3 (2nd ed), New York: Academic Press.

Lukas, J. R., Aigner, M., Blumer, R., Heinzl, H., & Mayr, R. (1994). Number and distribution of neuromuscular spindles in human extraocular muscles. *Investigative Ophthalmology and Visual Science, 35*, 4317–4327.

Mach, E. (1906/1959). *The analysis of sensations and the relation of the physical to the psychical.* (New York: Dover, 1959). From the first German ed., C. M. Williams (Trans.) and revised and supplemented from the fifth German ed. by Sydney Waterlow, M. A.

Mack, A., & Bachant, J. (1969). Perceived movement of the afterimage during eye movements. *Perception and Psychophysics, 6*, 379–384.

Mack, A., & Herman, E. (1972). A new illusion: The underestimation of distance during pursuit eye movements. *Perception and Psychophysics, 12*, 471–473.

MacKay, D. M. (1970). Elevation of the visual threshold by displacement of retinal image. *Nature, 225*, 90–92.

MacKenzie, W. (1843). On asthenopia or weak-sightedness. *Edinburgh Journal of Medicine and Surgery, 60*, 73–103.

Maddox, E. E. (1893). *The Clinical Use of Prisms and the Decentering of lenses* (2nd ed.). Bristol, England: John Wright and Co.

Magendie, F. (1822). Experiments on the functions of the spinal nerve roots. Translated and reprinted in R. H. Herrnstein & E. G. Boring (Eds.), *A sourcebook in the history of psychology* (1965, pp. 19–22). Cambridge, MA: Harvard University Press.

Marg, E. (1951). An investigation of voluntary as distinguished from reflex accommodation. *American Journal of Optometry and Archives of American Academy of Optometry, 28*, 347–356.

Mateeff, S., Hohnsbein, J., & Ehrenstein, W. A. (1990). Visual localization and estimation of extent of target motion during ocular pursuit: A common mechanism? *Perception, 19*, 459–469.

Matin, E., Clymer, B., & Matin, L. (1972). Metacontrast and saccadic suppression. *Science, 178*, 179–182.

Matin, L. (1976). A possible hybrid mechanism for modification of visual direction associated with eye movements–The paralyzed-eye experiment reconsidered. *Perception, 5*, 233–239.

Matin, L. (1986). Visual localization and eye movements (Chap. 20). In K. R. Boff, L. Kaufman, & J. P. Thomas (Eds.) *Handbook of perception and human performance: Vol 1. Sensory processes and perception.* New York: John Wiley and Sons.

Matin, L., & Fox, C. R. (1989). Visually perceived eye level and perceived elevation of objects: Linearly additive influences from visual field pitch and from gravity. *Vision Research, 29*, 315–324.

Matin, L., & Pearce, D. G. (1965). Visual perception of direction for stimuli flashed during voluntary saccadic eye movements. *Science, 148*, 1485–1488.

Matin, L., Picoult, E. Stevens, J. K., Edwards, M. W. Jr., Young, D., & MacArthur, R. (1982). Oculoparalytic illusion: Visual-field dependent spatial mislocalizations by humans partially paralyzed with curare. *Science, 216*, 198–201.

Matin, L., Stevens, J. K., & Picoult, E. (1983). Perceptual consequences of experimental extraocular muscle paralysis (chap. 14). In A. Hein and M. Jeannerod (Eds.), *Spatially oriented behavior.* New York: Springer-Verlag.

McBrien, N. A., & Millidot, M. (1987). A biometric investigation of late onset myopic eyes. *Acta Ophthalmologica, 65*, 461–468.

McBrien, N. A., & Millodot, M. (1988). Differences in adaptation of tonic accommodation with refractive state. *Investigative Ophthalmology and Visual Science, 29*, 460–469.

McLaughlin, S. C. (1967). Parametric adjustment in saccadic eye movements. *Perception and Psychophysics, 2*, 359–362.

McLaughlin, S. C., Kelly, M. J., Anderson, R. E., & Wenz, T. G. (1968). Localization of a peripheral target during parametric adjustment of saccadic eye movements. *Perception and Psychophysics, 4*, 45–48.

McLaughlin, S. C., & Webster, R. G. (1967). Changes in straight-ahead eye position during adaptation to wedge prisms. *Perception and Psychophysics, 2*, 37–44.

McLin, L. N. Jr., & Schor, C. M. (1988). Voluntary effort as a stimulus to accommodation and vergence. *Investigative Ophthalmology and Visual Science, 29*, 1739–1746.

Melvill Jones, G. (1970). Origin significance and amelioration of Coriolis' illusions from the semicircular canals: A non-mathematical appraisal. *Aerospace Medicine, 41*, 483–490.

Melvill Jones, G., Berthoz, A., & Segal, B. N. (1984). Adaptive modification of the vestibulo-ocular reflex by mental effort in darkness. *Experimental Brain Research, 56*, 149–153.

Merton, P. A. (1972). How we control the contraction of our muscles. *Scientific American, 226*, 30–37.

Meyer, C. H., Lasher, G. L., & Robinson, D. A. (1985). The upper limit of human smooth pursuit velocity. *Vision Research, 25*, 561–563.

Miles, F. A., & Eighmy, B. B. (1980). Long-term adaptive changes in primate vestibulo-ocular reflex. I. Behavioral observations. *Journal of Neurophysiology, 43*, 1406–1425.

Miles, F. A., Judge, S. J., & Optican, L. M. (1987). Optically induced changes in the couplings between vergence and accommodation. *Journal of Neuroscience, 7*, 2576–2589.

Miles, F. A., & Lisberger, S. G. (1981). The "error" signals subserving adaptive gain control in the primate vestibulo-ocular reflex. *Annals of the New York Academy of Sciences, 374*, 513–525.

Miller, E. F. (1962). Counterrolling of the human eyes produced by head tilt with respect to gravity. *Acta Laryngology, 54*, 479–501.

Miller, J. M. (1980). Information used by the perceptual and oculomotor systems regarding the amplitude of saccadic and pursuit eye movements. *Vision Research, 20*, 59–68.

Miller, R. J., Pigion, R. G., & Takahama, M. (1986). The effects of ingested alcohol on accommodative, fusional, and dark vergence. *Perception and Psychophysics, 39*, 25–31.

Milot, J. A., Jacob, J. L., Blanc, V. F., & Hardy, J. F. (1983). The oculocardiac reflex in strabismus surgery. *Canadian Journal of Ophthalmology, 18*, 314–317.

Money, K. E. (1970). Motion sickness. *Psychological Review, 50*, 1–39.

Money, K. E. (1990). Motion sickness and evolution. In G. H. Crampton, (Ed.), *Motion and Space Sickness*. Boca Raton, FL: CRC Press, Inc.

Money, K. E., & Cheung, B. S. (1983). Another function of the inner ear: Facilitation of the emetic response to poisons. *Aviation, Space, and Environmental Medicine, 54*, 208–211.

Money, K. E., Kirienko, N. M., Watt, D. G. D., Johnson, W. H., Markham, C. H., & Diamond, S. G. (1987). Ocular torsion in response to hypogravity. *Proceedings of the symposium on vestibular organs and altered force environments, NASA, Space Biomedical Research Institute Symposium* (pp. 61–67). Houston, October, 1987.

Money, K. E., & Myles, W. S. (1974). Heavy water nystagmus and effects of alcohol. *Nature, 247*, 404–405.

Mon-Williams, M., & Tresilian, J. R. (1999). Some recent studies on the extraretinal contribution to distance perception. *Perception, 17*, 609–621.

Morgan, C. L. (1978). Constancy of egocentric visual direction. *Perception and Psychophysics, 23*, 61–68.

Müller, G. E. (1916). Über das Äubertsche Phanomenon. *Zeitschrift fur sinnesphysiologie, 49*, 109–244.

Müller, J. (1826). *Beitrage zur vergleichenden Physiologie des Gesichtssinnes*, Leipzig: Cnobloch.

Muratore, R., & Zee, D. S. (1979). Pursuit after-nystagmus. *Vision Research, 19*, 1057–1059.

Nakayama, K., & Balliet, R. (1977). Listing's law, eye position sense, and perception of the vertical. *Vision Research, 17*, 453–457.

National Research Council Working Group on Myopia Prevalence and Progression. (1989). Washington, DC: National Academy Press.

Newton, I. (1730/1952). *Opticks, Query 15* (pp. 346–347). New York: Dover Publications (based on the 4th ed., London, 1730).

Normann, R. A. (1995). Visual neuroprosthetics – functional vision for the blind. *IEEE Engineering in Medicine and Biology, 14*, 77–83.

Ogle, K. N., & Prangen, A. de H. (1953). Observations on vertical divergences and hyperphorias. *AMA Archives of Ophthalmology, 49*, 313–334.

Ohmi, M., Howard, I. P., & Landolt, J. P. (1987). Circular vection as a function of foregound–background relationship. *Perception, 16*, 17–22.

Ono, H., & Gonda, G. (1978). Apparent movement, eye movement, and phoria when the viewing eyes alternate in viewing a stimulus. *Perception, 7*, 75–83.

Optican, L. M., & Robinson, D. A. (1980). Cerebellar-dependent adaptive control of the primate saccadic system. *Journal of Neurophysiology, 44*, 1058–1076.

O'Shea, R. P., McDonald, A. A., Cummings, A., Pearl, D., Sanderson, G., & Molterno, A. C. B. (1994). Interocular transfer of the movement aftereffect in central and peripheral vision of people with strabismus. *Investigative Ophthalmology and Visual Science, 35*, 313–317.

Owens, D. A., & Leibowitz, H. W. (1976). Oculomotor adjustments in darkness and the specific distance tendency. *Perception and Psychophysics, 20*, 2–9.

Owens, D. A., & Leibowitz, H. W. (1980). Accommodation, convergence, and distance perception in low illumination. *American Journal of Optometry and Physiological Optics, 57*, 540–550.

Owens, D. A., & Leibowitz, H. W. (1983). Perceptual and motor consequences of tonic vergence. Ch. 3 in C. M. Schor & K. J. Ciuffreda (Eds.), *Vergence eye movements: Basic and clinical aspects.* Boston: Butterworths.

Owens, D. A., & Reed, E. S. (1994). Seeing where we look. *Behavioral and Brain Sciences, 17*, 271–272.

Owens, D. A., & Wolf-Kelly, K. (1987). Near work, visual fatigue, and variations of oculomotor tonus. *Investigative Ophthalmology and Visual Science, 28*, 743–749.

Owens, R. L., & Higgins, K. E. (1983). Long term stability of the dark focus of accommodation. *American Journal of Optometry and Physiological Optics, 60*, 32–38.

Paap, K. R., & Ebenholtz, S. M. (1976). Perceptual consequences of potentiation in the extraocular muscles: An alternative explanation for adaptation to wedge prisms. *Journal of Experimental Psychology: Human Perception and Performance, 2*, 457–468.

Paap, K. R., & Ebenholtz, S. M. (1977). Concomitant direction and distance aftereffects of sustained convergence: A muscle potentiation explanation for eye-specific adaptation. *Perception and Psychophysics, 21*, 307–314.

Paige, G. D., & Tomko, D. L. (1991a). Eye movement responses to linear head motion in the squirrel monkey: I. Basic Characteristics. *Journal of Neurophysiology, 65*, 1183–1196.

Paige, G. D., & Tomko, D. L. (1991b). Eye movement responses to linear head motion in the squirrel monkey. II. Visual–vestibular interactions and kinematic considerations. *Journal of Neurophysiology, 65*, 1183–1196.

Park, K., & Shebilske, W. L. (1991). Phoria, Hering's laws and monocular perception of direction. *Journal of Experimental Psychology: Human Perception and Performance, 17*, 219–231.

Parker, D. E. (1980). The vestibular apparatus. *Scientific American, 243*, 120.

Parker, D. E., Reschke, M. F., Arrott, A. P., Homick, J. L., & Lichtenberg, B. K. (1985). Otolith tilt-translation reinterpretation following prolonged weightlessness: Implications for preflight training. *Aviation, Space, and Environmental Medicine, 56*, 601–606.

Pelz, J. B., & Hayhoe, M. M. (1995). The role of exocentric reference frames in the perception of visual direction. *Vision Research, 35*, 2267–2275.

Penfield, W., & Perot, P. (1963). A brain's record of auditory and visual experience. A final summary and discussion. *Brain, 86*, 595–596.

Perrone, J. A. (1992). Model for the computation of self-motion in biological systems. *Journal of the Optical Society of America A., 9*, 177–194.

Peters, R. S. (1965). *Brett's history of psychology* (pp. 610–627). Cambridge, MA: The MIT Press.

Pola, J., & Wyatt, H. J. (1989). The perception of target motion during smooth pursuit eye movements in the open-loop condition: Characteristics of retinal and extraretinal signals. *Vision Research, 29*, 471–483.

Porter, J. D., & Balaban, C. D. (1997). Connections between the vestibular nuclei and brain stem regions that mediate autonomic function in the rat. *Journal of Vestibular Research, 7*, 63–76.

Post, R. B., & Leibowitz, H. W. (1982). The effect of convergence on the vestibulo-ocular reflex and implications for perceived movement. *Vision Research, 22*, 461–465.

Post, R. B., & Leibowitz, H. W. (1985). A revised analysis of the role of efference in motion perception. *Perception, 14*, 631–643.

Post, R. B., & Lott, L. A. (1992). The relationship between vestibulo-ocular reflex plasticity and changes in apparent concomitant motion. *Vision Research, 32*, 89–96.

Post, R. B., Shupert, C. L., & Leibowitz, H. W. (1984). Implications of OKN suppression by smooth pursuit for induced motion. *Perception and Psychophysics, 36*, 493–498.

Provine, R. R., & Enoch, J. M. (1975). On voluntary ocular accommodation. *Perception and Psychophysics, 17*, 209–212.

Pyykko, I., Schalen, L., & Matsuoka, I. (1985). Transdermally administered scopolamine vs. dimenhydrinate. II. Effect on different types of nystagmus. *Acta Otolaryngologica* (Stockholm), *99*, 597–604.

Raphan, T., Matsuo, V., & Cohen, B. (1979). Velocity storage in the vestibulo-ocular reflex arc (VOR). *Experimental Brain Research, 35*, 229–248.

Rashbass, C., & Westheimer, G. (1961). Disjunctive eye movements. *Journal of Physiology* (London), *159*, 339–360.

Reason, J. T., & Brand, J. J. (1975). *Motion sickness.* New York: Academic Press.

Reichardt, W., & Poggio, T. (1976). Visual control of orientation behavior in the fly. *Quarterly Review of Biophysics, 9*, 311–375.

Riggs, L. A., Merton, P. A., & Morton, H. B. (1974). Suppression of visual phosphenes during saccadic eye movements. *Vision Research, 14*, 997–1011.

Rine, R. M., & Skavenski, A. A. (1997). Extraretinal eye position signals determine perceived target location when they conflict with visual cues. *Vision Research, 37*, 775–787.

Robinson, D. A. (1965). The mechanics of human smooth pursuit eye movements. *Journal of Physiology, 180*, 569–591.

Robinson, D. A. (1976). Adaptive gain control of vestibuloocular reflex by the cerebellum. *Journal of Neurophysiology, 39*, 954–969.

Rock, I. (1983). *The logic of perception.* Cambridge, MA: MIT Press.

Rock, I. (1990). The frame of reference. In I. Rock (Ed.), *The legacy of Solomon Asch: Essays in cognition and social psychology.* Hillsdale, NJ: Lawrence Erlbaum Associates.

Rock, I., & Ebenholtz, S. (1959). The relational determination of perceived size. *Psychological Review, 66*, 387–401.

Rock, I., & Ebenholtz, S. M. (1962). Stroboscopic movement based on change of phenomenal rather than retinal location. *American Journal of Psychology, 75*, 193–207.

Rock, I., Tauber, E. S., & Heller, D. (1965). Perception of stroboscopic movement: Evidence for its innate basis. *Science, 147*, 1050–1052.

Rogers, B., & Graham. M. (1982). Similarities between motion parallax and stereopsis in human depth perception. *Vision Research, 22*, 261–270.

Roll, J. P., & Roll, R. (1987). Kinaesthetic and motor effects of extraocular muscle vibration in man. In J. K. O'Ryan & A. Levy-Schoen (Eds.), *Eye movements: From physiology to cognition.* North-Holland: Elsevier Science.

Roll, R., Velay, J. L., & Roll, J. P. (1991). Eye and neck proprioceptive messages contribute to the spatial coding of retinal input in visually oriented activities. *Experimental Brain Research, 85*, 424–431.

Ron, S., Robinson, D. A., & Skavenski, A. A. (1972). Saccades and the quick phase of nystagmus. *Vision Research, 12,* 2015–2022.

Rose, D. (1999). The historical roots of the theories of local signs and labeled lines. *Perception, 28,* 675–685.

Rosenhall, U. (1972). Vestibular macular mapping in man. *Annals of Otolaryngology, 81*, 339–351.

Ross, H. E. (1975). *Behavior and perception in strange environments.* New York: Basic Books.

Royden, C. S., Crowell, J. A., & Banks, M. S. (1994). Estimating heading during eye movements. *Vision Research, 34,* 3197–3214.

Scheerer, E. (1987). Muscle sense and innervation feelings: A chapter in the history of perception and action. In H. Heuer & A. F. Saunders (Eds.), *Perspectives on perception and action.* Hillsdale, NJ: Erlbaum.

Schor, C. M. (1979). The relationship between fusional vergence eye movements and fixation disparity. *Vision Research, 19,* 1359–1367.

Schor, C. M., Johnson, C. A., & Post, R. B. (1984). Adaptation of tonic accommodation. *Ophthalmic and Physiological Optics, 4*, 133–137.

Schor, C. M., Lott, L. A., Pope, D., & Graham, A. D. (1999). Saccades reduce latency and increase velocity of ocular accommodation. *Vision Research, 39,* 3769–3795.

Schultheis, L. W., & Robinson, D. A. (1981). Directional plasticity of the vestibulo-ocular reflex in the cat. *New York Academy of Science, 374,* 504–512.

Scott, A. (1995). *Stairway to the mind.* New York: Copernicus, Springer-Verlag New York, Inc.

Searle, J. R. (1992). *The rediscovery of the mind.* Cambridge, MA: MIT Press.

Semmlow, J. L., Stark, L. Vandepol, C., & Nguyen, A. (1991). The relationship between ciliary muscle contraction and accommodative response in the presbyopic eye (pp. 245–253). In G. Obrecht, & L.W. Stark (Eds.). *Presbyopia research, from molecular biology to visual adaptation.* New York: Plenum Press.

Semmlow, J. L., Yuan, W., & Alvarez, T. L. (1998). Evidence for separate control of slow version and vergence eye movements: Support for Hering's Law. *Vision Research, 38,* 1145–1152.

Service, R. F. (1999). Neurones and silicon get intimate. *Science, 284,* 578–579.

Sharp, J. A., Trovst, B. T., Dell'Osso, L. F., & Daroff, R. B. (1975). Comparative velocities of different types of fast eye movements in man. *Investigative Ophthalmology, 14,* 689–692.

Shebilske, W. L. (1976). Extraretinal information in corrective saccades and inflow vs. outflow theories of visual direction constancy. *Vision Research, 16,* 621–628.

Shebilske, W. L. (1981). Visual direction illusions in everyday situations: Implications for sensorimotor and ecological theories. (Part II, Chap. 3). In D. F. Fisher, R. A. Monty, & J. W. Senders (Eds.), *Eye movements: Cognition and visual perception.* Hillsdale, NJ: Erlbaum.

Shebilske, W. L. (1986). Baseball batters support an ecological efference mediation theory of natural event perception. *Acta Psychologica, 63,* 117–131.

Shebilske, W. L. (1987). An ecological efference mediation theory of natural event perception. In H. Heuer & A. F. Sanders (Eds.), *Perspectives on perception and action* (pp. 195–213). Hillsdale, NJ: Earlbaum.

Shebilske, W. L., & Karmiohl, C. M. (1978). Illusory visual direction during and after backward head tilts. *Perception and Psychophysics, 24,* 543–545.

Shebilske, W. L., Karmiohl, & Proffitt, D. R. (1983). Induced esophoric shifts in eye convergence and illusory distance in reduced and structured viewing conditions. *Journal of Experimental Psychology: Human Perception and Performance, 9,* 270–277.

Shelhamer, M., Robinson, D. A., & Tan, H. S. (1992). Context specific adaptation of the gain of the vestibulo-ocular reflex in humans. *Vestibular Research, 2,* 89–96.

Sherrington, C. S. (1893). Further experimental note on the correlation of action of anatgonistic muscles. *Proceedings of the Royal Society, B53,* 407–420.

Sherrington, C. S. (1918). Observations on the sensual role of the proprioceptive nerve supply of the extrinsic eye muscles. *Brain, 41,* 332–343.

Sireteanu, R., Lagreze, W.-D., & Constantinescu, D. H. (1993). Distortion in two-dimensional visual space perception in strabismic observers. *Vision Research, 33,* 677–690.

Skavenski, A. A. (1972). Inflow as a source of extraretinal eye position information. *Vision Research, 12,* 221–229.

Skavenski, A. A., Haddad, G., & Steinman, R. M. (1972). The extraretinal signal for the visual perception of direction. *Perception and Psychophysics, 11,* 287–290.

Smith, S. T., Curthoys, I. S., & Moore, S. T. (1995). The human ocular torsion position response during yaw angular acceleration. *Vision Research, 35,* 2045–2055.

Sperry, R. W. (1950) Neural basis of the spontaneous optokinetic response produced by visual inversion. *Journal of Comparative Physiology and Psychology, 34,* 482–489.

Sperry, R. (1985). Some effects of disconnecting the cerebral hemispheres, pp. 372–380. In P. H. Abelson, E. Butz, & S. H. Snyder (Eds.), *Neuroscience.* Washington, DC: AAAS.

Stark, L., & Bridgeman, B. (1983). Role of corollary discharge in space constancy. *Perception and Psychophysics, 34,* 371–380.

Stark, L., & Ellis, S. R. (1981). Scanpaths revisited: cognitive models direct active looking. Chapter IV.1 in D. F. Fisher, R. A. Monty, & J. W. Senders (Eds.), *Eye movements: cognition and visual perception.* Hillsdale, NJ: Lawrence Erlbaum Associates.

Steinbach, M. J. (1976). Pursuing the perceptual rather than the retinal stimulus. *Vision Research, 16,* 1371–1376.

Steinbach, M. J., & Money, K. E. (1973). Eye movements of the owl. *Vision Research, 13,* 889–891.

Steinbach, M. J. (1987). Proprioceptive knowledge of eye position. *Vision Research, 27,* 1737–1744.

Steinman, R. M., Cunitz, R. J., Timberlake, G. T., & Herman, M. (1967). Voluntary control of microsaccades during maintained monocular fixation. *Science, 155,* 1577–1579.

Stenstrom, S. (1948). Investigation of the variation and the correlation of the optical elements of human eyes (D. Woolf, Trans.). *American Journal of Optometry and Archives of American Academy of Optometry, Monograph No. 58,* 1–71.

Sterling, T. D., Bering, E. A. Jr., Pollack, S. V., & Vaughan, H. G. (Eds.). (1971). *Visual prosthesis.* New York: Academic Press.

Stern, R. M., Hu, S. A., Anderson, R. B., Leibowitz, H. W., & Koch, K. L. (1990). The effects of fixation and restricted visual field on vection-induced motion sickness. *Aviation, Space, and Environmental Medicine, 61,* 712–715.

Stevens, G. T. (1887). The anomalies of the ocular muscles. *Archives of Ophthalmology, 16,* 149–176.

Stevens, J. K., Emerson, R. C., Gerstein, G. L., Kallos, T., Neufeld, G. R., Nichols, C. W., & Rosenquist, A. C. (1976). Paralysis of the awake human: Visual perceptions. *Vision Research, 16,* 93–98.

Stoper, A. E., & Cohen, M. M. (1986). Judgments of eye level in light and in darkness. *Perception and Psychophysics, 40,* 311–316.

Tamar, H. (1972). *Principles of sensory physiology.* Springfield, II: Charles C. Thomas.

Tan, R. K. T., & O'Leary, D. J. (1986). Stability of the accommodative dark focus after periods of maintained accommodation. *Investigative Ophthalmology and Visual Science, 27,* 1414–1417.

Tononi, G., & Edelman, G. M. (1998). Consciousness and complexity. *Science, 282,* 1846–1851.

Treisman, M. (1977). Motion sickness: An evolutionary hypothesis. *Science, 197,* 493–495.

Turvey, M. T., & Solomon, J. (1984). Visually perceiving distance. A comment on Shebilske, Karmiohl, & Proffitt (1983). *Journal of Experimental Psychology: Human Perception and Performance, 10,* 449–454.

Tyrrell, R. A., & Leibowitz, H. W. (1990). The relation of vergence effort to reports of visual fatigue following prolonged near work. *Human Factors, 32,* 341–357.

Vaegan (1976). The position of random autokinetic movement and the physiological position of rest are frequently stable and identical. *Perception and Psychophysics, 19,* 240–245.

van Damme, W., & Brenner, E. (1997). The distance used for scaling disparities is the same as the one used for scaling retinal size. *Vision Research, 37,* 757–764.

Van den Berg, A. V., & Collewijn, H. (1986). Human smooth pursuit: Effects of stimulus extent and of spatial and temporal constraints of the pursuit trajectory. *Vision Research, 26,* 1209–1222.

Van den Berg, A. V., & Collewijn, H. (1988). Directional asymmetries of human optokinetic nystagmus. *Experimental Brain Research, 70,* 597–604.

Van der Steen, J., Tamminga, E. P., & Collewijn, H. (1983). A comparison of oculomotor pursuit of a target in circular real, beta or sigma motion. *Vision Research, 23,* 1655–1661.

Van Die, G., & Collewijn, H. (1982). Optokinetic nystagmus in man. *Human Neurobiology, 1,* 111–119.

van Ee, R., & Erkelens, C. J. (1996). Stability of binocular depth perception with moving head and eyes. *Vision Research, 36,* 3827–3842.

Velay, J. L., Allin, F., & Bouquerel, A. (1997). Motor and perceptual responses to horizontal and vertical eye vibration in humans. *Vision Research, 37,* 2631–2638.

Velay, J. L., Roll, R., Lennerstrand, G., & Roll, J. P. (1994). Eye proprioception and visual localization in humans: Influence of ocular dominance and visual context. *Vision Research, 34,* 2169–2176.

Vieth, G. U. A. (1818). Über die Richtung der Augen. *Annalen der Physik, Leipzig, 28,* 233–253.

Vision research. A national plan: 1994–1998. A report of the National Advisory Eye Council, National Eye Institute, U.S. Department of Health and Human Services, Public Health Service, NIH Publications No. 95–3186.

Vogel, H., & Kass, J. R. (1986). European vestibular experiments on the Spacelab 1 mission: 7. Ocular counterrolling measurements pre- and post-flight. *Experimental Brain Research, 64,* 284–290.

von Bekesy, G. (1967). *Sensory inhibition*. Princeton, NJ: Princeton University Press.

von Holst, E. (1954). Relations between the central nervous system and the peripheral organs. *British Journal of Animal Behavior, 2*, 89–94.

von Holst, E., & Mittelstaedt, H. (1950). The principle of reafference: Interactions between the central nervous system and the peripheral organs. In P. C. Dodwell (Ed. & Trans.), *Perceptual processing: Stimulus equivalence and pattern recognition* (1971). New York: Appleton-Century-Crofts.

von Noorden, G. K. (1980). *Burian-von Noorden's binocular vision and ocular motility: Theory and management of strabismus* (2nd ed.). St. Louis: The C. V. Mosby Co.

Wade, N. J. (1998). Light and sight since antiquity. *Perception, 27*, 637–670.

Wade, S. W., & Curthoys, I. S. (1997). The effect of ocular torsional position on perception of the roll-tilt of visual stimuli. *Vision Research, 37*, 1071–1078.

Waespe, W., & Henn, V. (1977). Neuronal activity in the vestibular nuclei of the alert monkey during vestibular and optokinetic stimulation. *Experimental Brain Research, 27*, 523–538.

Wallach, H. (1976), *Hans Wallach on perception*. (Chap. 10, Perceptual learning). New York: Quadrangle/The New York Times Book Co.

Wallach, H., Moore, M., & Davidson, L. (1963). Modification of stereoscopic depth perception. *American Journal of Psychology, 76*, 191–204.

Wallach, H., Schuman, P., & O'Leary, A. (1981). The effect of prolonged practice of pursuit eye movement. *Perception and Psychophysics, 30*, 533–539.

Wallach, H., & Zuckerman, C. (1963). The constancy of stereoscopic depth. *American Journal of Psychology, 76*, 404–412.

Walls, G. L. (1942). Eye movements and the fovea. In *The vertebrate eye* (Chap. 10). Bloomfield Hills, RI: Cranbrook Institute of Science.

Walls, G. L. (1962). The evolutionary history of eye movements. *Vision Research, 2*, 69–80.

Watkins, L. R., & Mayer, D. J. (1985). Organization of endogenous opiate and nonopiate pain control systems, pp. 355–371. In P. H. Abelson, E. Butz, & S. H. Snyder (Eds.). *Neuroscience*. Washington, DC: AAAS.

Webster's universal dictionary of the English language (1936). New York: The World Syndicate Publishing Co.

Welch, R. B. (1978). *Perceptual modification: Adapting to altered sensory environments*. New York: Academic Press.

Westheimer, G., & Mitchell, A. M. (1956). Eye movement responses to convergence stimuli. *AMA. Arch. Ophthalmology, 55*, 848–856.

Westheimer, G., & Mitchell, D. E. (1969). The sensory stimulus for disjunctive eye movements. *Vision Research, 9*, 749–755.

Whiteside, T. C. D., Graybiel, A., & Niven, J. I. (1965). Visual illusions of movement. *Brain, 88*, 193–210.

Whitteridge, D. (1959). The effect of stimulation of intrafusal muscle fibers on sensitivity to stretch of extraocular muscle spindles. *Quarterly Journal of Experimental Physiology, 44*, 385–393.

Wyatt, H. J., & Pola, J. (1983). Smooth pursuit eye movements under open-loop and closed-loop conditions. *Vision Research, 23*, 1121–1131.

Wyatt, H. J., & Pola, J. (1987). Smooth eye movements with step-ramp stimuli: The influence of attention and stimulus extent. *Vision Research, 27*, 1565–1580.

Yessenow, M. D. (1972, May). An investigation of the relationship between nystagmus eye movements and the oculogyral illusion. Paper presented at the annual scientific meeting of the Aerospace Medical Association. Bal Harbour, FL.

Young, F. A. (1963). The effect of restricted visual space on the refractive error of the young monkey eye. *Investigative Ophthalmology, 2*, 571–577.

Young, L. R., Oman, C. M., Watt, D. G. D., Money, K. E., & Lichtenberg, B. K. (1984). Spatial orientation in weightlessness and readaptation to Earth's gravity. *Science, 225*, 205–208.

Subject Index

A and E effects, 95–97
  see also apparent vertical
accommodation: as adaptive control
    system, 38
  and ciliary muscle, 35
  reflexive, 36
  resting level of, 37
  vergence stimulated, 43
  voluntary control of, 39
  and zonules, 35–36
  see also CA/C ratio
  see also hysteresis effects
acuity: 20/20, 161
  and legal blindness, 161
amblyopia, strabismic: and anomalous
    correspondence, 168
  and suppression scotomas, 168
ametropia, 159
apparent horizon, 88
  and gaze, 89–90
  in pitchroom, 88–90
  and pitchbox, 88–90
apparent vertical: and apparent elevation, 94
  with backward pitch, 95–96
  in frontal plane, 97–101
  and ocular torsion, 98–101
  see also A and E effects
aqueous humor: and canal of Schlemm, 36
  and ciliary process, 36

pressure of in glaucoma, 168
Aristotle, and apparent visual direction, 118
asthenopia, 170–171
astigmatism, 166
  and aniseikonia, 166
Aubert-Fleischl effect, 117
axial length, 165
  distribution of, 165
  and refractive error, 165
  and x-rays, 165

body tilt: in 1-g field, 100
  in centrifuge, 100

CA/C ratio, 43
  adaptation of, 43
chromatophores: and eye color, 40
compensation theory, 125–129
  and exafference, 126
  and position constancy, 126–127
  and reafference, 126
  and voluntary innervation, 127–129
  see also oculogyral illusion
conscious states: as direct reality, 4–5
  in simulated brain, 6–7
Coriolis maneuver: and cross coupled
    rotation, 152
  and nausea, 151
  and nystagmus, 150

201

Author Index

Reference list pages are in bold.